Henry White Warren

Recreations in Astronomy

Henry White Warren

Recreations in Astronomy

ISBN/EAN: 9783337395704

Printed in Europe, USA, Canada, Australia, Japan

Cover: Foto ©berggeist007 / pixelio.de

More available books at **www.hansebooks.com**

RECREATIONS IN ASTRONOMY

WITH

DIRECTIONS FOR PRACTICAL EXPERIMENTS AND TELESCOPIC WORK

BY

HENRY WHITE WARREN, D.D.

AUTHOR OF "SIGHTS AND INSIGHTS; OR, KNOWLEDGE BY TRAVEL," ETC.

WITH EIGHTY-THREE ILLUSTRATIONS AND MAPS OF STARS

NEW YORK

CHAUTAUQUA PRESS

C. L. S. C. DEPARTMENT

805 BROADWAY

1886

PREFACE.

ALL sciences are making an advance, but Astronomy is moving at the double-quick. Since the principles of this science were settled by Copernicus, four hundred years ago, it has never had to beat a retreat. It is re-written not to correct material errors, but to incorporate new discoveries.

Once Astronomy treated mostly of tides, seasons, and telescopic aspects of the planets; now these are only primary matters. Once it considered stars as mere fixed points of light; now it studies them as suns, determines their age, size, color, movements, chemical constitution, and the revolution of their planets. Once it considered space as empty; now it knows that every cubic inch of it quivers with greater intensity of force than that which is visible in Niagara. Every inch of surface that can be conceived of between suns is more wave-tossed than the ocean in a storm.

The invention of the telescope constituted one era in Astronomy; its perfection in our day, another; and the discoveries of the spectroscope a third—no less important than either of the others.

While nearly all men are prevented from practical experimentation in these high realms of knowledge, few have so little leisure as to be debarred from intelligently enjoying the results of the investigations of others.

This book has been written not only to reveal some of the highest achievements of the human mind, but also to let the heavens declare the glory of the Divine Mind. In the author's judgment, there is no gulf that separates science and religion, nor any conflict where they stand together. And it is fervently hoped that any one who comes to a better knowledge of God's works through reading this book, may thereby come to a more intimate knowledge of the Worker.

I take great pleasure in acknowledging my indebtedness to J. M. Van Vleck, LL.D., of the U. S. Nautical Almanac staff, and Professor of Astronomy at the Wesleyan University, for inspecting some of the more important chapters; to Dr. S. S. White, of Philadelphia, for telescopic advantages; to Professor Henry Draper, for furnishing, in advance of publication, a photograph of the sun's corona in 1878; and to the excellent work on "Popular Astronomy," by Professor Simon Newcomb, LL.D., Professor U. S. Naval Observatory, for some of the most recent information, and for the use of the unequalled engravings of Jupiter, Saturn, and the great nebula of Orion.

CONTENTS.

ILLUSTRATIONS.

I.

CREATIVE PROCESSES.

"In the beginning God created the heaven and the earth. And the earth was without form, and void; and darkness was upon the face of the deep."—*Genesis* i. 1, 2.

1

"Not to the domes, where crumbling arch and column
    Attest the feebleness of mortal hand,
But to that fane, most catholic and solemn,
    Which God hath planned,—
To that cathedral, boundless as our wonder,
    Whose quenchless lamps the sun and moon supply,
Its choir the winds and waves, its organ thunder,
    Its dome the sky."   HORACE SMITH

"The heavens are a point from the pen of His perfection;
The world is a rose-bud from the bower of His beauty;
The sun is a spark from the light of His wisdom;
And the sky a bubble on the sea of His power."
        SIR W. JONES

RECREATIONS IN ASTRONOMY.

I.

CREATIVE PROCESSES.

DURING all the ages there has been one bright and glittering page of loftiest wisdom unrolled before the eye of man. That this page may be read in every part, man's whole world turns him before it. This motion apparently changes the eternally stable stars into a moving panorama, but it is only so in appearance. The sky is a vast, immovable dial-plate of "that clock whose pendulum ticks ages instead of seconds," and whose time is eternity. The moon moves among the illuminated figures, traversing the dial quickly, like a second-hand, once a month. The sun, like a minute-hand, goes over the dial once a year. Various planets stand for hour-hands, moving over the dial in various periods reaching up to one hundred and sixty-four years; while the earth, like a ship of exploration, sails the infinite azure, bearing the observers to different points where they may investigate the infinite problems of this mighty machinery.

This dial not only shows present movements, but it keeps the history of uncounted ages past ready to be read backward in proper order; and it has glorious volumes of prophecy, revealing the far-off future to any man who is able to look thereon, break the seals, and read the record. Glowing stars are the alphabet of this lofty page. They combine to form words. Meteors, rainbows, auroras, shifting groups of stars, make pictures vast and significant as the armies, angels, and falling stars in the Revelation of St. John—changing and progressive pictures of infinite wisdom and power.

Men have not yet advanced as far as those who saw the pictures John describes, and hence the panorama is not understood. That continuous speech that day after day uttereth is not heard; the knowledge that night after night showeth is not seen; and the invisible things of God from the creation of the world, even his eternal power and Godhead, clearly discoverable from things that are made, are not apprehended.

The greatest triumphs of men's minds have been in astronomy—and ever must be. We have not learned its alphabet yet. We read only easy lessons, with as many mistakes as happy guesses. But in time we shall know all the letters, become familiar with the combinations, be apt at their interpretation, and will read with facility the lessons of wisdom and power that are written on the earth, blazoned in the skies, and pictured by the flowers below and the rainbows above.

In order to know how worlds move and develop, we must create them; we must go back to their beginning, give their endowment of forces, and study the laws of their unfolding. This we can easily do by that faculty wherein man is likest his Father, a creative imagination. God creates and embodies; we create, but it remains in thought only. But the creation is as bright, strong, clear, enduring, and real, as if it were embodied. Every one of us would make worlds enough to crush us, if we could embody as well as create. Our ambition would outrun our wisdom. Let us come into the high and ecstatic frame of mind which Shakspeare calls frenzy, in the exigencies of his verse, when

"The poet's eye, in a fine frenzy rolling,
Doth glance from heaven to earth, from earth to heaven;
And, as imagination bodies forth
The forms of things unknown, the poet's pen
Turns them to shapes, and gives to airy nothing
A local habitation and a name."

In the supremacy of our creative imagination let us make empty space, in order that we may therein build up a new universe. Let us wave the wand of our power, so that all created things disappear. There is no world under our feet, no radiant clouds, no blazing sun, no silver moon, nor twinkling stars. We look up, there is no light; down, through immeasurable abysses, there is no form; all about, and there is no sound or sign of being—nothing save utter silence, utter darkness. It cannot be endured. Creation is a necessity of mind— even of the Divine mind.

We will now, by imagination, create a monster world, every atom of which shall be dowered with the single power of attraction. Every particle shall reach out its friendly hand, and there shall be a drawing together of every particle in existence. The laws governing this attraction shall be two. When these particles are associated together, the attraction shall be in proportion to the mass. A given mass will pull twice

as much as one of half the size, because there is twice
as much to pull.   And a given mass will be pulled
twice as much as one half as large, because there is
twice as much to be pulled.   A man who weighed one
hundred and fifty pounds on the earth might weigh a
ton and a half on a body as large as the sun.   That
shall be one law of attraction; and the other shall be
that masses attract inversely as the square of distances
between them.   Absence shall affect friendships that
have a material basis.   If a body like the earth pulls a
man one hundred and fifty pounds at the surface, or
four thousand miles from the centre, it will pull the
same man one-fourth as much at twice the distance,
one-sixteenth as much at four times the distance.   That
is, he will weigh by a spring balance thirty-seven and a
half pounds at eight thousand miles from the centre,
and nine pounds six ounces at sixteen thousand miles
from the centre, and he will weigh or be pulled by the
earth $\frac{1}{34}$ of a pound at the distance of the moon.   But
the moon would be large enough and near enough to
pull twenty-four pounds on the same man, so the earth
could not draw him away.   Thus the two laws of attraction of gravitation are—1, *Gravity is proportioned
to the quantity of matter;* and 2, *The force of gravity
varies inversely as the square of the distance from the
centre of the attracting body.*

The original form of matter is gas.   Almost as I
write comes the announcement that Mr. Lockyer has
proved that all the so-called primary elements of matter are only so many different sized molecules of one
original substance — hydrogen.   Whether that is true
or not, let us now create all the hydrogen we can

imagine, either in differently sized masses or in combination with other substances. There it is! We cannot measure its bulk; we cannot fly around it in any recordable eons of time. It has boundaries, to be sure, for we are finite, but we cannot measure them. Let it alone, now; leave it to itself. What follows? It is dowered simply with attraction. The vast mass begins to shrink, the outer portions are drawn inward. They rush and swirl in vast cyclones, thousands of miles in extent. The centre grows compact, heat is evolved by impact, as will be explained in Chapter II. Dull red light begins to look like coming dawn. Centuries go by; contraction goes on; light blazes in insufferable brightness; tornadoes, whirlpools, and tempests scarcely signify anything as applied to such tumultuous tossing.

There hangs the only world in existence; it hangs in empty space. It has no tendency to rise; none to fall; none to move at all in any direction. It seethes and flames, and holds itself together by attractive power, and that is all the force with which we have endowed it.

Leave it there alone, and withdraw millions of miles into space: it looks smaller and smaller. We lose sight of those distinctive spires of flame, those terrible movements. It only gives an even effulgence, a steady unflickering light. Turn one quarter round. Still we see our world, but it is at one side.

Now in front, in the utter darkness, suddenly create another world of the same size, and at the same distance from you. There they stand—two huge, lone bodies, in empty space. But we created them dowered with attraction. Each instantly feels the drawing influence of the other. They are mutually attractive, and begin to

move toward each other. They hasten along an unde-
viating straight line. Their speed quickens at every
mile. The attraction increases every moment. They
fly swift as thought. They dash their flaming, seething
foreheads together.

And now we have one world again. It is twice as
large as before, that is all the difference. There is no
variety, neither any motion ; just simple flame, and nothing to be warmed thereby. Are our creative powers
exhausted by this effort ?

No, we will create another world, and add another
power to it that shall keep them apart. That power

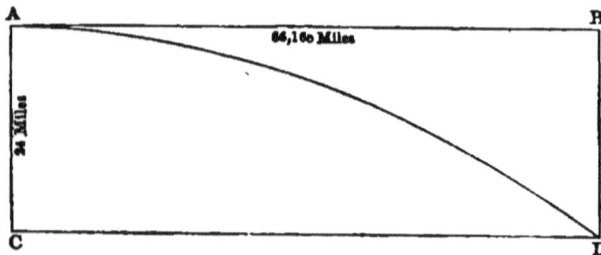

Fig. 1.—Orbit A D, resulting from attraction, A C, and projectile force, A B.

shall be what is called the force of inertia, which is
literally no power at all ; it is an inability to originate
or change motion. If a body is at rest, inertia is that
quality by which it will forever remain so, unless acted
upon by some force from without; and if a body is in
motion, it will continue on at the same speed, in a
straight line, forever, unless it is quickened, retarded,
or turned from its path by some other force. Suppose
our newly created sun is 860,000 miles in diameter. Go
away 92,500,000 miles and create an earth eight thousand miles in diameter. It instantly feels the attractive power of the sun drawing it to itself twenty-

four miles the first hour. Now, just as it starts, give this earth a push in a line at right angles with line of fall to the sun, that shall send it 66,168 miles every hour thereafter. It obeys both forces. The result is that the world moves constantly forward at the same speed by its inertia from that first push, and attraction momentarily draws it from its straight line, so that the new world circles round the other to the starting-point. Continuing under the operation of both forces, the worlds can never come together or fly apart.

They circle about each other as long as these forces endure; for the first world does not stand still and the second do all the going; both revolve around the centre of gravity common to both. In case the worlds are equal in mass, they will both take the same orbit around a central stationary point, midway between the two. In case their mass be as one to eighty-one, as in the case of the earth and the moon, the centre of gravity around which both turn will be $\frac{1}{82}$ of the distance from the earth's centre to the moon's centre. This brings the central point around which both worlds swing just inside the surface of the earth. It is like an apple attached by a string, and swung around the hand; the hand moves a little, the apple very much.

Thus the problem of two revolving bodies is readily comprehended. The two bodies lie in easy beds, and swing obedient to constant forces. When another body, however, is introduced, with its varying attraction, first on one and then on the other, complications are introduced that only the most masterly minds can follow. Introduce a dozen or a million bodies, and complications arise that only Omniscience can unravel.

Let the hand swing an apple by an elastic cord. When the apple falls toward the earth it feels another force besides that derived from the hand, which greatly lengthens the elastic cord. To tear it away from the earth's attraction, and make it rise, requires additional force, and hence the string is lengthened; but when it passes over the hand the earth attracts it downward, and the string is very much shortened: so the moon, held by an elastic cord, swings around the earth. From its extreme distance from the earth, at A, Fig. 2, it rushes with increasing speed nearly a quarter of a million of miles toward the sun, feeling its attraction increase with every mile until it reaches B; then it is retarded in its speed, by the same attraction, as it climbs back its quarter of a million of miles away from the sun, in defiance of its power, to C. All the while the invisible elastic force of the earth is unweariedly maintained; and though the moon's distances vary over a range of 31,355 miles, the moon is always in a determinable place. A simple revolution of one world about another in a circular orbit would be a problem of easy solution. It would always be at the same distance from its centre, and going with the same velocity. But there are over sixty causes that interfere with such a simple orbit in the case of the moon, all of which causes and their disturbances must be considered in calculating such a simple matter as an eclipse, or predicting the moon's place as the sailors' guide. One of the most puzzling of the irregularities

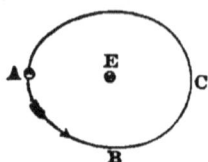

of our night-wandering orb has just been explained by
Professor Hansen, of Gotha, as a curious result of the
attraction of Venus.

Take a single instance of the perturbations of Jupiter
and Saturn which can be rendered evident. The times
of orbital revolution of Saturn and Jupiter are nearly as
five to two. Suppose the orbits
of the planets to be, as in Fig. 3,
both ellipses, but not necessarily
equally distant in all parts. The
planets are as near as possible
at 1, 1. Drawn toward each other by mutual attraction, Jupiter's orbit bends outward, and
Saturn's becomes more nearly

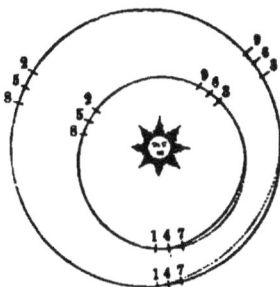

Fig. 3.—Changes of orbit by
mutual attraction.

straight, as shown by the dotted lines. A partial correction of this difficulty immediately follows. As Jupiter moves on ahead of Saturn it is held back—retarded
in its orbit by that body; and Saturn is hastened in its
orbit by the attraction of Jupiter. Now greater speed
means a straighter orbit. A rifle-ball flies nearer in a
straight line than a thrown stone. A greater velocity
given to a whirled ball pulls the elastic cord far enough
to give the ball a larger orbit. Hence, being hastened,
Saturn stretches out nearer its proper orbit, and, retarded, Jupiter approaches the smaller curve that is its true
orbit.

But if they were always to meet at this point, as they
would if Jupiter made two revolutions to Saturn's one,
it would be disastrous. In reality, when Saturn has
gone around two-thirds of its orbit to 2, Jupiter will
have gone once and two-thirds around and overtaken

Saturn; and they will be near again, be drawn together, hastened, and retarded, as before; their next conjunction would be at 3, 3, etc.

Now, if they always made their conjunction at points
equally distant, or at thirds of their orbits, it would cause
a series of increasing deviations; for Jupiter would
be constantly swelling his orbit at three points, and
Saturn increasingly contracting his orbit at the same
points. Disaster would be easily foretold. But as their
times of orbital revolutions are not exactly in the ratio
of five and two, their points of conjunction slowly travel
around the orbit, till, in a period of nine hundred years,
the starting-point is again reached, and the perturbations have mutually corrected one another.

For example, the total attractive effect of one planet
on the other for 450 years is to quicken its speed. The
effect for the next 450 years is to retard. The place of
Saturn, when all the retardations have accumulated for
450 years, is one degree behind what it is computed if
they are not considered; and 450 years later it will be
one degree before its computed place—a perturbation
of two degrees. When a bullet is a little heavier or
ragged on one side, it will constantly swerve in that direction. The spiral groove in the rifle, of one turn in
forty-five feet, turns the disturbing weight or raggedness
from side to side—makes one error correct another, and
so the ball flies straight to the bull's-eye. So the place of
Jupiter and Saturn, though further complicated by four
moons in the case of Jupiter, and eight in the case of
Saturn, and also by perturbations caused by other planets, can be calculated with exceeding nicety.

The difficulties would be greatly increased if the orbits of Saturn and Jupiter, instead of being 400,000,000 miles apart, were interlaced. Yet there are the orbits of two hundred and fifty asteroids so interlaced that, if they were made of wire, no one could be lifted without raising the whole net-work of them. Nevertheless, all these swift chariots of the sky race along the course of their intermingling tracks as securely as if they were each guided by an intelligent mind. *They are guided by an intelligent mind and an almighty arm.*

Still more complicated is the question of the mutual attractions of all the planets. Lagrange has been able to show, by a mathematical genius that seems little short of omniscience in his single department of knowledge, that there is a discovered system of oscillations, affecting the entire planetary system, the periods of which are immensely long. The number of these oscillations is equal to that of all the planets, and their periods range from 50,000 to 2,000,000 years.

Looking into the open page of the starry heavens we see double stars, the constituent parts of which must revolve around a centre common to them both, or rush to a common ruin. Eagerly we look to see if they revolve, and beholding them in the very act, we conclude, not groundlessly, that the same great law of gravitation holds good in distant stellar spaces, and that there the same sufficient mind plans, and the same sufficient power directs and controls all movements in harmony and security.

When we come to the perturbations caused by the mutual attractions of the sun, nine planets, twenty moons, two hundred and fifty asteroids, millions of comets, and innumerable meteoric bodies swarming in space, and when we add to all these, that belong to one solar system, the attractions of all the systems of the other suns that sparkle on a brilliant winter night, we are compelled to say, " As high as the heavens are above the earth, so high above our thoughts and ways must be the thoughts and ways of Him who comprehends and directs them all."

II.

CREATIVE PROGRESS.

"And God said, Let there be light, and there was light." — *Genesis* i., 3.

"God is light."—*1 John*, i., 5.

"Hail! holy light, offspring of Heaven first born,
Or of the eternal, co-eternal beam,
May I express thee unblamed? since God is light,
And never but in unapproached light
Dwelt from eternity, dwelt then in thee,
Bright effluence of bright essence increate."

MILTON.

"A million torches lighted by Thy hand
Wander unwearied through the blue abyss:
They own Thy power, accomplish Thy command,
All gay with life, all eloquent with bliss.
What shall we call them?   Piles of crystal light—
A glorious company of golden streams—
Lamps of celestial ether burning bright—
Suns lighting systems with their joyous beams?
But Thou to these art as the noon to night."

DERZHAVIN, trans. by BOWRING.

II.

CREATIVE PROGRESS.

WORLDS would be very imperfect and useless when simply endowed with attraction and inertia, if no time were allowed for these forces to work out their legitimate results. We want something more than swirling seas of attracted gases, something more than compacted rocks. We look for soil, verdure, a paradise of beauty, animal life, and immortal minds. Let us go on with the process.

Light is the child of force, and the child, like its father, is full of power. We dowered our created world with but a single quality—a force of attraction. It not only had attraction for its own material substance, but sent out an all-pervasive attraction into space. By the force of condensation it flamed like a sun, and not only lighted its own substance, but it filled all space with the luminous outgoings of its power. A world may be limited, but its influence cannot; its body may have bounds, but its soul is infinite. Everywhere is its manifestation as real, power as effective, presence as actual, as at the central point. He that studies ponderable bodies alone is not studying the universe, only its skeleton. Skeletons are somewhat interesting in themselves, but far more so when covered with flesh, flushed with beauty, and inspired with soul. The universe has bones.

flesh, beauty, soul, and all is one. It can be understood only by a study of all its parts, and by tracing effect to cause.

But how can condensation cause light? Power cannot be quiet. The mighty locomotive trembles with its own energy. A smitten piece of iron has all its infinitesimal atoms set in vehement commotion; they surge back and forth among themselves, like the waves of a storm-blown lake. Heat is a mode of motion. A heated body commences a vigorous vibration among its particles, and communicates these vibrations to the surrounding air and ether. When these vibrations reach 396,000,000,000,000 per second, the human eye, fitted to be affected by that number, discerns the emitted undulations, and the object seems to glow with a dull red light; becoming hotter, the vibrations increase in rapidity. When they reach 765,000,000,000,000 per second the color becomes violet, and the eye can observe them no farther. Between these numbers are those of different rapidities, which affect the eye—as orange, yellow, green, blue, indigo, in an almost infinite number of shades—according to the sensitiveness of the eye.

We now see how our dark immensity of attractive atoms can become luminous. A force of compression results in vibrations within, communicated to the ether, discerned by the eye. Illustrations are numerous. If we suddenly push a piston into a cylinder of brass, the force produces heat enough to set fire to an inflammable substance within. Strike a half-inch cube of iron a moderate blow and it becomes warm; a sufficient blow, and its vibrations become quick enough to be seen —it is red-hot. Attach a thermometer to an extended arm of a whirling wheel; drive it against the air five hundred feet per second, the mercury rises 16°. The earth goes 98,000 feet per second, or one thousand miles a minute. If it come to an aerolite or mass of metallic rock, or even a cloudlet of gas, standing still in space, its contact with our air evolves 600,000° of heat. And when the meteor comes toward the world twenty-six miles a second, the heat would become proportionally greater if the meteor could abide it, and not be consumed in fervent heat. It vanishes almost as soon as seen. If there were meteoric masses enough lying in our path, our sky would blaze with myriads of flashes of light. Enough have been seen to enable a person to read by them at night. If a sufficient number were present, we should miss their individual flashes as they blend their separate fires in one sea of insufferable glory. The sun is 326,800 as heavy as our planet; its attraction proportionally greater; the aerolites more numerous; and hence an infinite hail of stones, small masses and little worlds, makes ceaseless trails of light, whose individuality is lost in one dazzling sea of glory.

On the 1st day of September, 1859, two astronomers, independently of each other, saw a sudden brightening on the surface of the sun. Probably two large meteoric masses were travelling side by side at two or three hundred miles per second, and striking the sun's atmosphere, suddenly blazed into light bright enough to be seen on the intolerable light of the photosphere as a background. The earth responded to this new cause of brilliance and heat in the sun. Vivid auroras appeared, not only at the north and south poles, but even where such spectacles are seldom seen. The electromagnetic disturbances were more distinctly marked. " In many places the telegraphic wires struck work. In Washington and Philadelphia the electric signalmen received severe electric shocks; at a station in Norway the telegraphic apparatus was set fire to; and at Boston a flame of fire followed the pen of Bain's electric telegraph." There is the best of reason for believing that a continuous succession of such bodies might have gone far toward rendering the earth uncomfortable as a place of residence.

Of course, the same result of heat and light would follow from compression, if a body had the power of contraction in itself. We endowed every particle of our gas, myriads of miles in extent, with an attraction for every other particle. It immediately compressed itself into a light-giving body, which flamed out through the interstellar spaces, flushing all the celestial regions with exuberant light.

But heat exerts a repellent force among particles, and soon an equilibrium is reached, for there comes a time when the contracting body can contract no farther. But heat and light radiate away into cold space, then contraction goes on evolving more light, and so the suns flame on through the millions of years unquenched. It is estimated that the contraction of our sun, from filling immensity of space to its present size, could not afford heat enough to last more than 18,000,000 years, and that its contraction from its present density (that of a swamp) to such rock as that of which our earth is composed, could supply heat enough for 17,000,000 years longer. But the far-seeing mind of man knows a time must come when the present force of attraction shall have produced all the heat it can, and a new force of attraction must be added, or the sun itself will become cold as a cinder, dead as a burned-out char.

Since light and heat are the product of such enormous cosmic forces, they must partake of their nature, and be force. So they are. The sun has long arms, and they are full of unconquerable strength ninety-two millions, or any other number of millions, of miles away. All this light and heat comes through space that is 200° below zero, through utter darkness, and appears only on the earth. So the gas is darkness in the underground pipes, but light at the burner. So the electric power is unfelt by the cable in the bosom of the deep, but is expressive of thought and feeling at the end. Having found the cause of light, we will commence a study of its qualities and powers.

Light is the astronomer's necessity. When the sublime word was uttered, " Let there be light!" the study of astronomy was made possible. Man can gather but little of it with his eye; so he takes a lens twenty-six inches in diameter, and bends all the light that passes through it to a focus, then magnifies the image and takes it into his eye. Or he takes a mirror, six feet in diameter, so hollowed in the middle as to reflect all the rays falling upon it to one point, and makes this larger eye fill his own with light. By this larger light-gathering he discerns things for which the light falling on his pupil one-fifth of an inch in diameter would not be sufficient. We never have seen any sun or stars; we have only seen the light that left them a few minutes or years ago, more or less. Light is the aërial sprite that carries our measuring-rods across the infinite

spaces; light spreads out the history of that far-off beginning; brings us the measure of stars a thousand times
brighter than our sun; takes up into itself evidences of
the very constitutional elements of the far-off suns, and
spreads them at our feet. It is of such capacity that
the Divine nature, looking for an expression of its own
omnipotence, omniscience, and power of revelation, was
content to say, "God is Light." We shall need all our
delicacy of analysis and measurement when we seek to
determine the activities of matter so fine and near to
spirit as light.

We first seek the velocity of light. In Fig. 4 the
earth is 92,500,000 miles from the sun at E; Jupiter
is 480,000,000 miles from the sun at J. It has four

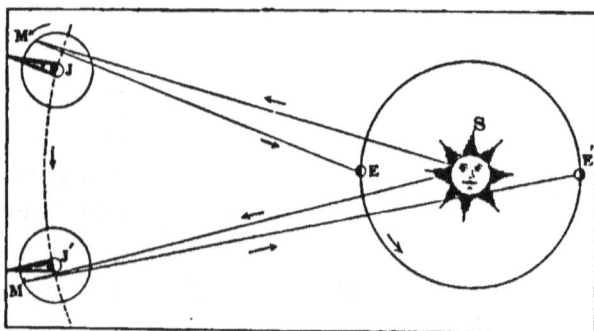

Fig. 4.—Velocity of Light measured by Eclipses of Jupiter's Moons.

moons: the inner one goes around the central body
in forty-two hours, and is eclipsed at every revolution.
The light that went out from the sun to M ceases to be
reflected back to the earth by the intervention of the planet Jupiter. We know to a second when these eclipses
take place, and they can be seen with a small telescope.
But when the earth is on the opposite side of the sun

from Jupiter, at E', these eclipses at J' take place sixteen and a half minutes too late. What is the reason? Is the celestial chronometry getting deranged? No, indeed; these great worlds swing never an inch out of place, nor a second out of time. By going to the other side of the sun the earth is 185,000,000 miles farther from Jupiter, and the light that brings the intelligence of that eclipse consumes the extra time in going over the extra distance. Divide one by the other and we get the velocity, 186,868 miles per second. That is probably correct to within a thousand miles. Methods of measurement by the toothed wheel of Fizeau confirm this result. Suppose the wheel, Fig. 5, to have one thousand teeth, making five revolutions to the second. Five thousand flashes of light each second will dart out. Let each flash travel nine miles to a mirror

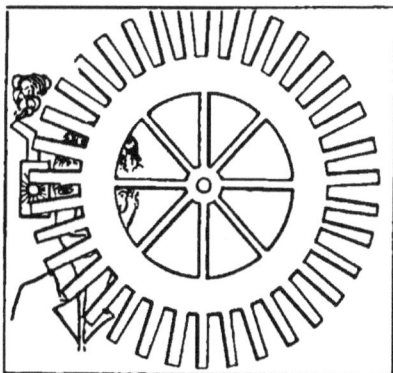

Fig. 5.—Measuring the Velocity of Light.

and return. If it goes that distance in $\frac{1}{10000}$ of a second, or at the rate of 180,000 miles a second, the next tooth will have arrived before the eye, and each returning ray be cut off. Hasten the revolutions a little, and the next notch will then admit the ray, on its return, that went out of each previous notch: the eighteen miles having been traversed meanwhile. The result of experiments by Lieut. Michelson, as given Sept., 1879, is 186,380 miles per second, which cannot be fifty miles from correct.

When we take instantaneous photographs by the exposure of the sensitive plate $\frac{1}{20000}$ part of a second, a stream of light nine miles long dashes in upon the plate in that very brief period of time.

The highest velocity we can give a rifle-ball is 2000 feet a second, the next second it is only 1500 feet, and soon it comes to rest. We cannot compact force enough behind a bit of lead to keep it flying. But light flies unweariedly and without diminution of speed. When it has come from the sun in eight minutes, Alpha Centauri in three years, Polaris in forty-five years, other stars in one thousand, its wings are in nowise fatigued, nor is the rapidity of its flight slackened in the least.

It is not the transactions of to-day that we read in the heavens, but it is history, some of it older than the time of Adam. Those stars may have been smitten out of existence decades of centuries ago, but their poured-out light is yet flooding the heavens.

It can go both ways at once in the same place, without interference. We see the light reflected from the new moon to the earth ; reflected back from the house-tops, fields, and waters of earth, to the moon again, and from the moon to us once more—three times in opposite directions, in the same place, without interference, and thus we see " the old moon in the arms of the new."

Constitution of Light.

Light was once supposed to be corpuscular, or consisting of transmitted particles. It is now known to be the result of undulations in ether. Reference has been made to the minuteness of these undulations. Their velocity is equally wonderful. Put a prism of glass into a ray of light coming into a dark room, and it is instautly turned out of its course, some parts more and some less, according to the number of vibrations, and appears as the seven colors on different parts of the screen. Fig. 6 shows the arrangement of colors, and the number of millions of millions of vibrations per second of each.

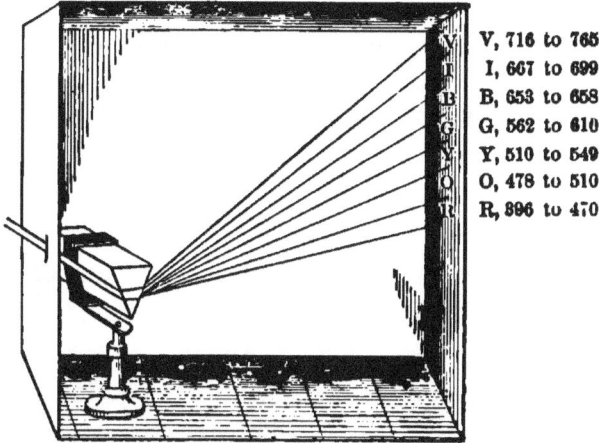

V,	716 to 765
I,	667 to 699
B,	653 to 658
G,	562 to 610
Y,	510 to 549
O,	478 to 510
R,	396 to 470

Fig. 6.—White Light resolved into Colors.

But the different divisions we call colors are not colors in themselves at all, but simply a different number of vibrations. Color is all in the eye. Violet has in different places from 716 to 765,000,000,000,000 of vibrations per second; red has, in different places, from 396 to 470,000,000,000,000 vibrations per second. None of these in any sense are color, but affect the eye differently, and we call these different effects color. They are simply various velocities of vibration. An object, like one kind of stripe in our flag, which absorbs all kinds of vibrations except those between 396 and 470,000,000,-000,000, and reflects those, appears red to us. The field for the stars absorbs and destroys all but those vibrations numbering about 653,000,000,000,000 of vibra-

2

tions per second. A color is a constant creation. Light makes momentary color in the flag. Drake might have written, in the continuous present as well as in the past,

> " Freedom mingles with its gorgeous dyes
> The milky baldrick of the skies,
> And stripes its pure celestial white
> With streakings of the morning light."

Every little pansy, tender as fancy, pearled with eva-nescent dew, fresh as a new creation of sunbeams, has power to suppress in one part of its petals all vibrations we call red, in another those we call yellow, and pur-ple, and reflect each of these in other parts of the same tender petal. "Pansies are for thoughts," even more thoughts than poor Ophelia knew. A cloud of smoke that is dense enough to absorb all the faster and weaker vibrations, leaving only the stronger to come through, will show the sun as red; because the vibrations that produce the impression we have so named are the only ones that have vigor enough to get through. It is like an army charging upon a fortress. Under the deadly fire and fearful obstructions six-sevenths go down, but one-seventh comes through with the glory of victory upon its face.

Light comes in undulations to the eye, as tones of sound to the ear. Must not light also sing? The lowest tone we can hear is made by 16.5 vibrations of air per second; the highest, so shrill and "fine that nothing lives 'twixt it and silence," is made by 38,000 vibrations per second. Between these extremes lie eleven octaves; C of the G clef having $258\frac{2}{3}$ vibrations to the second, and its octave above $517\frac{1}{3}$. Not that sound vibrations cease at 38,000, but our organs are not fitted to hear beyond those limitations. If our ears were delicate enough, we could hear even up to the almost infinite vibrations of light. In one of those semi-inspirations we find in Shakspeare's works, he says—

> "'There's not the smallest orb which thou beholdest,
> But in his motion like an angel sings,
> Still quiring to the young-eyed cherubim.
> Such harmony is in immortal souls;
> But, whilst this muddy vesture of decay
> Doth grossly close it in, we cannot hear it."

And that older poetry which is always highest truth says, "The morning stars sing together." We misconstrued another passage which we could not understand, and did not dare translate as it was written, till science crept up to a perception of the truth that had been standing there for ages, waiting a mind that could take it in. Now we read as it is written—"Thou makest the out-goings of the morning and evening to sing." Were our senses fine enough, we could hear the separate keynote of every individual star. Stars differ in glory and in power, and so in the volume and pitch of their song. Were our hearing sensitive enough, we could hear not only the separate key-notes but the infinite swelling harmony of these myriad stars of the sky, as they pour their mighty tide of united anthems in the ear of God:

> "In reason's ear they all rejoice,
> And utter forth a glorious voice.
> Forever singing, as they shine,
> The hand that made us is divine."

This music is not monotonous. Stars draw near each other, and make a light that is unapproachable by mortals; then the music swells beyond our ability to endure. They recede far away, making a light so dim that the music dies away, so near to silence that only spirits can perceive it. No wonder God rejoices in his works. They pour into his ear one ceaseless tide of rapturous song.

Our senses are limited—we have only five, but there is room for many more. Some time we shall be taken out of "this muddy vesture of decay," no longer see the universe through crevices of our prison-house, but shall range through wider fields, explore deeper mysteries, and discover new worlds, hints of which have never yet been blown across the wide Atlantic that rolls between them and men abiding in the flesh.

Chemistry of Suns revealed by Light.

When we examine the assemblage of colors spread from the white ray of sunlight, we do not find red simple red, yellow yellow, etc., but there is a vast number of fine microscopic lines of various lengths, parallel—here near together, there far apart, always the same number and the same relative distance, when the same light and prism are used. What new alphabets to new realms of knowledge are these! Remember, that what we call colors are only various numbers of vibrations of ether. Remember, that every little group in the infinite variety of these vibrations may be affected differently from every other group. One number of these is bent by the prism to where we see what we call the violet, another number to the place we call red. All of the vibrations are destroyed when they strike a surface we call black. A part of them are destroyed when they

strike a substance we call colored. The rest are reflected, and give the impression of color. In one place on
the flag of our nation all vibrations are destroyed except the red; in another, all but the blue. Perhaps on
that other gorgeous flag, not of our country but of our
sun, the flag we call the solar spectrum, all vibrations
are destroyed where these dark lines appear. Perhaps
this effect is not produced by the surface upon which
the rays fall, but by some specific substance in the sun.
This is just the truth. Light passing through vapor of
sodium has the vibrations that would fall on two narrow lines in the yellow utterly destroyed, leaving two
black spaces. Light passing through vapor of burning
iron has some four hundred numbers or kinds of vibrations destroyed, leaving that number of black lines;
but if the salt or iron be glowing gas, in the source of
the light itself the same lines are bright instead of
dark.

Thus we have brought to our doors a readable record of the very substances composing every world hot
enough to shine by its own light. Thus, while our flag
means all we have of liberty, free as the winds that kiss.
it, and bright as the stars that shine in it, the flag of the
sun means all that it is in constituent elements, all that
it is in condition.

We find in our sun many substances known to exist
in the earth, and some that we had not discovered when
the sun wrote their names, or rather made their mark,
in the spectrum. Thus, also, we find that Betelguese
and Algol are without any perceivable indications of
hydrogen, and Sirius has it in abundance. What a sense
of acquaintanceship it gives us to look up and recognize

the stars whose very substance we know! If we were transported thither, or beyond, we should not be altogether strangers in an unknown realm.

But the stars differ in their constituent elements; every ray that flashes from them bears in its very being proofs of what they are. Hence the eye of Omniscience, seeing a ray of light anywhere in the universe, though gone from its source a thousand years, would be able to tell from what orb it originally came.

Creative Force of Light.

Just above the color vibrations of the unbraided sunbeam, above the violet, which is the highest number our eyes can detect, is a chemical force; it works the changes on the glass plate in photography; it transfigures the dark, cold soil into woody fibre, green leaf, downy rose petals, luscious fruit, and far pervasive odor; it flushes the wide acres of the prairie with grass and flowers, fills the valleys with trees, and covers the hills with corn, a single blade of which all the power of man could not make.

This power is also fit and able to survive. The engineer Stephenson once asked Dr. Buckland, "What is the power that drives that train?" pointing to one thundering by. "Well, I suppose it is one of your big engines." "But what drives the engine?" "Oh, very likely a canny Newcastle driver." "No, sir," said the engineer, "it is sunshine." The doctor was too dull to take it in. Let us see if we can trace such an evident effect to that distant cause. Ages ago the warm sunshine, falling on the scarcely lifted hills of Pennsylvania, caused the reedy vegetation to grow along the banks of shallow seas, accumulated vast amounts of this vegetation, sunk it beneath the sea, roofed it over with sand, compacted the sand into rock, and changed this vegetable matter—the products of the sunshine—into coal; and when it was ready, lifted it once more, all garnered for the use of men, roofed over with mighty mountains. We mine the coal, bring out the heat, raise the steam, drive the train, so that in the ultimate analyses it is sunshine that drives the train. These great beds of coal are nothing but condensed sunshine—the sun's great force, through ages gone, preserved for our use to-day. And it is so full of force that a piece of coal that will weigh three pounds (as big as a large pair of fists) has as much power in it as the average man puts into a day's work. Three tons of coal will pump as much water or shovel as much sand as the average man will pump or shovel in a lifetime; so that if a man proposes to do nothing but work with his muscles, he had better dig three tons of coal and set that to do his work and then die, because his work will be better done, and without any cost for the maintenance of the doer.

Come down below the color vibrations, and we shall find that those which are too infrequent to be visible, manifest as heat. Naturally there will be as many different kinds of heat as tints of color, because there is as great a range of numbers of vibration. It is our privilege to sift them apart and sort them over, and find what kinds are best adapted to our various uses.

Take an electric lamp, giving a strong beam of light and heat, and with a plano-convex lens gather it into a single beam and direct it upon a thermometer, twenty feet away, that is made of glass and filled with air. The

expansion or contraction of this air will indicate the
varying amounts of heat. Watch your air-thermometer,
on which the beam of heat is pouring, for the result.
There is none. And yet there is a strong current of
heat there. Put another kind of test of heat beyond
it and it appears; coat the air-thermometer with a bit
of black cloth, and that will absorb heat and reveal it.
But why not at first? Because the glass lens stops all
the heat that can affect glass. The twenty feet of air
absorbs all the heat that affects air, and no kind of heat
is left to affect an instrument made of glass and air;
but there are kinds of heat enough to affect instruments
made of other things.

A very strong current of heat may be sent right
through the heart of a block of ice without melting the
ice at all or cooling off the heat in the least. It is done
in this way: Send the beam of heat through water in
a glass trough, and this absorbs all the heat that can
affect water or ice, getting itself hot, and leaving all
other kinds of heat to go through the ice beyond; and
appropriate tests show that as much heat comes out on
the other side as goes in on this side, and it does not
melt the ice at all. Gunpowder may be exploded by
heat sent through ice. Dr. Kane, years ago, made this
experiment. He was coming down from the north,
and fell in with some Esquimaux, whom he was anxious to conciliate. He said to the old wizard of the
tribe, "I am a wizard; I can bring the sun down out
of the heavens with a piece of ice." That was a good
deal to say in a country where there was so little sun.
"So," he writes, "I took my hatchet, chipped a small
piece of ice into the form of a double-convex lens,

smoothed it with my warm hands, held it up to the sun, and, as the old man was blind, I kindly burned a blister on the back of his hand to show him I could do it."

These are simple illustrations of the various kinds of heat. The best furnace or stove ever invented consumes fifteen times as much fuel to produce a given amount of heat as the furnace in our bodies consumes to produce a similar amount. We lay in our supplies of carbon at the breakfast, dinner, and supper table, and keep ourselves warm by economically burning it with the oxygen we breathe.

Heat associated with light has very different qualities from that which is not. Sunlight melts ice in the middle, bottom, and top at once. Ice in the spring-time is honey-combed throughout. A piece of ice set in the summer sunshine crumbles into separate crystals. Dark heat only melts the surface.

Nearly all the heat of the sun passes through glass without hinderance; but take heat from white-hot platinum and only seventy-six per cent. of it goes through glass, twenty-four per cent. being so constituted that it cannot pass with facility. Of heat from copper at 752° only six per cent. can go through glass, the other ninety-four per cent. being absorbed by it.

The heat of the sunbeam goes through glass without any hinderance whatever. It streams into the room as freely as if there were no glass there. But what if the furnace or stove heat went through glass with equal facility? We might as well try to heat our rooms with the window-panes all out, and the blast of winter sweeping through them.

The heat of the sun, by its intense vibrations, comes

to the earth dowered with a power which pierces the
miles of our atmosphere, but if our air were as pervious .
to the heat of the earth, this heat would fly away every
night, and our temperature would go down to 200° below zero. This heat comes with the light, and then,
dissociated from it, the number of its vibrations lessened, it is robbed of its power to get away, and remains
to work its beneficent ends for our good.

Worlds that are so distant as to receive only $\frac{1}{1000}$ of
the heat we enjoy, may have atmospheres that retain it
all. Indeed it is probable that Mars, that receives but
one-quarter as much heat as the earth, has a temperature as high as ours. The poet drew on his imagination
when he wrote :

> " Who there inhabit must have other powers,
> Juices, and veins, and sense and life than ours ;
> One moment's cold like theirs would pierce the bone,
> Freeze the heart's-blood, and turn us all to stone."

The power that journeys along the celestial spaces
in the flashing sunshine is beyond our comprehension.
It accomplishes with ease what man strives in vain to
do with all his strength. At West Point there are some
links of a chain that was stretched across the river to
prevent British ships from ascending ; these links were
made of two-and-a-quarter-inch iron. A powerful locomotive might tug in vain at one of them and not stretch
it the thousandth part of an inch. But the heat of a
single gas-burner, that glows with the preserved sunlight of other ages, when suitably applied to the link,
stretches it with ease ; such enormous power has a little
heat. There is a certain iron bridge across the Thames
at London, resting on arches. The warm sunshine, act-

ing upon the iron, stations its particles farther and farther apart. Since the bottom cannot give way the arches must rise in the middle. As they become longer they lift the whole bridge, and all the thundering locomotives and miles of goods-trains cannot bring that bridge down again until the power of the sunshine has been withdrawn. There is Bunker Hill Monument, thirty-two feet square at the base, with an elevation of two hundred and twenty feet. The sunshine of every summer's day takes hold of that mighty pile of granite with its aërial fingers, lengthens the side affected, and bends the whole great mass as easily as one would bend a whipstock. A few years ago we hung a plummet from the top of this monument to the bottom. At 9 A.M. it began to move toward the west; at noon it swung round toward the north; in the afternoon it went east of where it first was, and in the night it settled back to its original place.

The sunshine says to the sea, held in the grasp of gravitation, "Rise from your bed! Let millions of tons of water fly on the wings of the viewless air, hundreds of miles to the distant mountains, and pour there those millions of tons that shall refresh a whole continent, and shall gather in rivers fitted to bear the commerce and the navies of nations." Gravitation says, "I will hold every particle of this ocean as near the centre of the earth as I can." Sunshine speaks with its word of power, and says, "Up and away!" And in the wreathing mists of morning these myriads of tons rise in the air, fly away hundreds of miles, and supply all the Niagaras, Mississippis and Amazons of earth. The sun says to the earth, wrapped in the mantle of winter,

"Bloom again;" and the snows melt, the ice retires, and vegetation breaks forth, birds sing, and spring is about us.

Thus it is evident that every force is constitutionally arranged to be overcome by a higher, and all by the highest. Gravitation of earth naturally and legitimately yields to the power of the sun's heat, and then the waters fly into the clouds. It as naturally and legitimately yields to the power of mind, and the waters of the Red Sea are divided and stand " upright as an heap." Water naturally bursts into flame when a bit of potassium is thrown into it, and as naturally when Elijah calls the right kind of fire from above. What seems a miracle, and in contravention of law, is only the constitutional exercise of higher force over forces organized to be swayed. If law were perfectly rigid, there could be but one force; but many grades exist from cohesion to mind and spirit. The highest forces are meant to have victory, and thus give the highest order and perfectness.

Across the astronomic spaces reach all these powers, making creation a perpetual process rather than a single act. It almost seems as if light, in its varied capacities, were the embodiment of God's creative power; as if, having said, "Let there be light," he need do nothing else, but allow it to carry forward the creative processes to the end of time. It was Newton, one of the earliest and most acute investigators in this study of light, who said, " I seem to have wandered on the shore of Truth's great ocean, and to have gathered a few pebbles more beautiful than common; but the vast ocean itself rolls before me undiscovered and unexplored."

EXPERIMENTS WITH LIGHT.

A light set in a room is seen from every place; hence light streams in every possible direction. If put in the centre of a hollow sphere, every point of the surface will be equally illumined. If put in a sphere of twice the diameter, the same light will fall on all the larger surface. The surfaces of spheres are as the squares of their diameters; hence, in the larger sphere the surface is illumined only one-quarter as much as the smaller. The same is true of large and small rooms. In Fig. 7 it is ap-

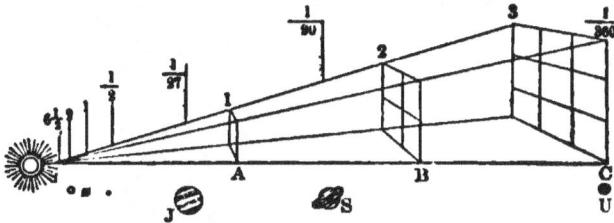

Fig. 7.

parent that the light that falls on the first square is spread, at twice the distance, over the second square, which is four times as large, and at three times the distance over nine times the surface. The varying amount of light received by each planet is also shown in fractions above each world, the amount received by the earth being 1.

Fig. 8.—Measuring Intensities of Lights.

The intensity of light is easily measured. Let two lights of different brightness, as in Fig. 8, cast shadows on the same screen. Arrange them as to distance so that both shadows shall be equally dark. Let them fall side by side, and study them carefully. Measure the respective distances. Suppose one is twenty inches, the other forty. Light varies as the square of the distance: the square of 20 is 400, of 40 is 1600. Divide 1600 by 400, and the result is that one light is four times as bright as the other.

Light can be handled, directed, and bent, as well as iron bars. Darken a room and admit a beam of sunlight through a shutter, or a ray of lamp-light through the key-hole. If there is dust in the room it will be observed that light goes in straight lines. Because of this men are able to arrange houses and trees in rows, the hunter aims his rifle correctly, and the astronomer projects straight lines to infinity. Take a hand-mirror, or bet-

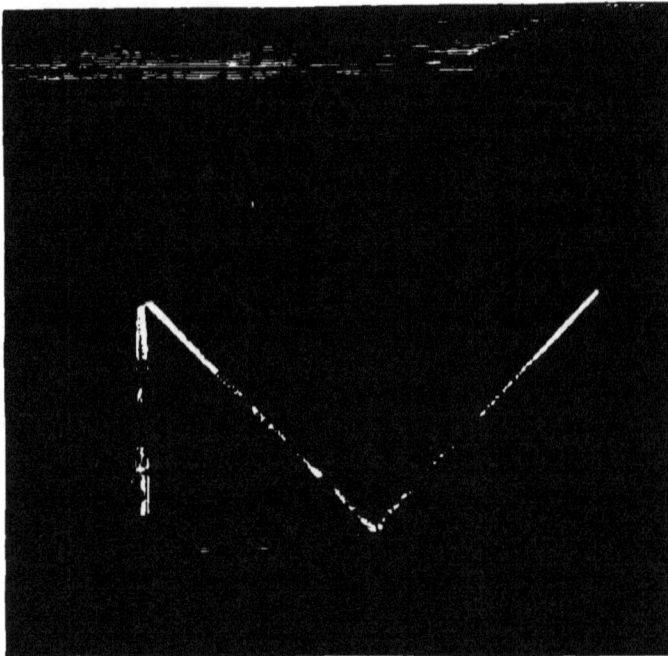

Fig. 9.—Reflection and Diffusion of Light.

ter, a piece of glass coated on one side with black varnish, and you can send your ray anywhere. By using two mirrors, or having an assistant and using several, you can cause a ray of light to turn as many corners as you please. I once saw Mr. Tyndall send a ray into a glass jar filled with smoke (Fig. 9). Admitting a slender ray through a small hole in a card over the mouth, one ray appeared; removing the cover, the whole jar was luminous; as the smoke disappeared in spots cavities of dark-ness appeared. Turn the same ray into a tumbler of water, it becomes faintly visible; stir into it a teaspoonful of milk, then turn in the ray of sunlight, and it glows like a lamp, illuminating the whole room. These experiments show how the straight rays of the sun are diffused in every direction over the earth.

Set a small light near one edge of a mirror; then, by putting the eye near the opposite edge, you see almost as many flames as you please from the multiplied reflections. How can this be accounted for?

Into your beam of sunlight, admitted through a half-inch hole, put the mirror at an oblique angle; you can arrange it so as to throw half a dozen bright spots on the opposite wall.

In Fig. 10 the sunbeam enters at A, and, striking the mirror *m* at *a*, is partly reflected to 1 on the wall, and partly enters the glass, passes

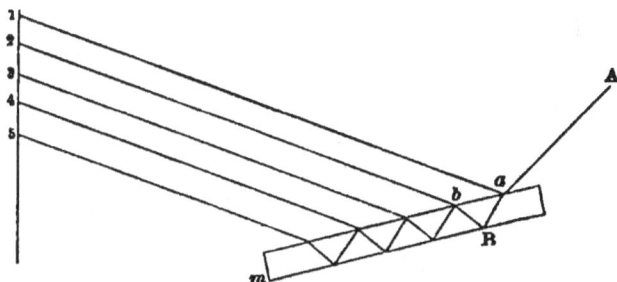

Fig. 10.—Manifold Reflections.

through to the silvered back at B, and is totally reflected to *b*, where it again divides, some of it going to the wall at 2, and the rest, continuing to make the same reflections and divisions, causes spots 3, 4, 5, etc. The brightest spot is at No. 2, because the silvered glass at B is the best reflector and has the most light.

When the discovery of the moons of Mars was announced in 1877, it was also widely published that they could be seen by a mirror. Of course this is impossible. The point of light mistaken for the moon in this secondary reflection was caused by holding the mirror in an oblique position.

Take a small piece of mirror, say an inch in surface, and putting under it three little pellets of wax, putty, or clay, set it on the wrist, with one of the pellets on the pulse. Hold the mirror steadily in the beam of light, and the frequency and prominence of each pulse-beat will be indicated by the tossing spot of light on the wall. If the operator becomes excited the fact will be evident to all observers.

Place a coin in a basin (Fig. 11), and set it so that the rim will conceal the coin from the eye. Pour in water, and the coin will appear

Fig. 11.

to rise into sight. When light passes from a medium of one density to a medium of another, its direction is changed. Thus a stick in water seems bent. Ships below the horizon are sometimes seen above, because of the different density of the layers of air.

Thus light coming from the interstellar spaces, and entering our atmosphere, is bent down more and more by its increasing density. The effect is greatest when the sun or star is near the horizon, none at all in the zenith. This brings the object into view before it is risen. Allowance for this displacement is made in all delicate astronomical observations.

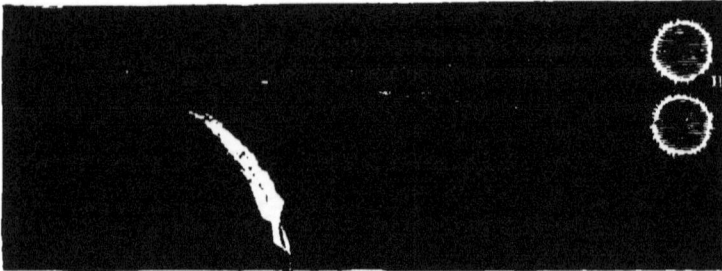

Fig. 12.—Atmospherical Refraction.

Notice on the floor the shadow of the window-frames. The glass of almost every window is so bent as to turn the sunlight aside enough to obliterate some of the shadows or increase their thickness.

DECOMPOSITION OF LIGHT.

Admit the sunbeam through a slit one inch long and one-twentieth of an inch wide. Pass it through a prism. Either purchase one or make it of three plain pieces of glass one and a half inch wide by six inches long, fastened together in triangular shape—fasten the edges with hot wax and fill it with water; then on a screen or wall you will have the colors of the rainbow, not merely seven but seventy, if your eyes are sharp enough.

Take a bit of red paper that matches the red color of the spectrum. Move it along the line of colors toward the violet. In the orange it is dark, in the yellow darker, in the green and all beyond, black. That is because there are no more red rays to be reflected by it. So a green object is true to its color only in the green rays, and black elsewhere. All these colors may be recombined by a second prism into white light.

III.

ASTRONOMICAL INSTRUMENTS.

"The eyes of the Lord are in every place."—*Proverbs* xv. 3

"Man, having one kind of an eye given him by his Maker, proceeds to construct two other kinds. He makes one that magnifies invisible objects thousands of times, so that a dull razor-edge appears as thick as three fingers, until the amazing beauty of color and form in infinitesimal objects is entrancingly apparent, and he knows that God's care of least things is infinite. Then he makes the other kind four or six feet in diameter, and penetrates the immensities of space thousands of times beyond where his natural eye can pierce, until he sees that God's immensities of worlds are infinite also."—BISHOP FOSTER.

III.

THE TELESCOPE.

FREQUENT allusion has been made in the previous chapter to discovered results. It is necessary to understand more clearly the process by which such results have been obtained. Some astronomical instruments are of the simplest character, some most delicate and complex. When a man smokes a piece of glass, in order to see an eclipse of the sun, he makes a simple instrument. Ferguson, lying on his back and slipping beads on a string at a certain distance above his eye, measured the relative distances of the stars. The use of more complex instruments commenced when Galileo applied the telescope to the heavens. He cannot be said to have invented the telescope, but he certainly constructed his own without a pattern, and used it to good purpose. It consists of a lens, O B (Fig. 13), which

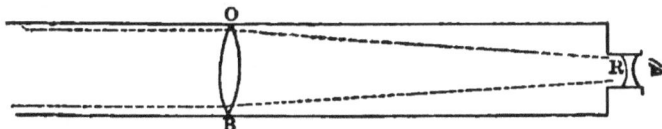

Fig. 13.—Refracting Telescope.

acts as a multiple prism to bend all the rays to one point at R. Place the eye there, and it receives as much light as if it were as large as the lens O B. The rays, however, are convergent, and the point difficult to find. Hence there is placed at R a concave lens, passing through which the rays emerge in parallel lines, and are received by the eye. Opera-glasses are made upon precisely this principle to-day, because they can be made conveniently short.

If, instead of a concave lens at R, converting the converging rays into parallel ones, we place a convex or magnifying lens, the minute image is enlarged as much as an object seems diminished when the telescope is reversed. This is the grand principle of the refracting telescope. Difficulties innumerable arise as we attempt to enlarge the instruments. These have been overcome, one after another, until we have now Struve's admirable glass of 30 inches, and an assurance of the speedy completion of the 36-inch lens for the Lick Observatory.

The Reflecting Telescope.

This is the only kind of instrument differing radically from the refracting one already described. It receives the light in a concave mirror, M (Fig. 14), which

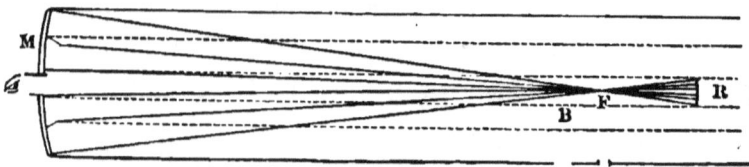

Fig. 14.—Reflecting Telescope.

reflects it to the focus F, producing the same result as the lens of the refracting telescope. At B a mirror may be placed obliquely, reflecting the image at right angles to the eye, outside the tube, in which case it is called the Newtonian telescope; or a mirror at R may be placed perpendicularly, and send the rays through an opening in the mirror at M. This form is called the Gregorian telescope. Or the mirror M may be slightly inclined to the coming rays, so as to bring the point F entirely outside the tube, in which case it is called the Herschelian telescope. In either case the image may be magnified, as in the refracting telescope.

Reflecting telescopes are made of all sizes, up to the Cyclopean eye of the one constructed by Lord Rosse, which is six feet in diameter. The form of instrument to be preferred depends on the use to which it is to be put. The loss of light in passing through glass lenses is about two-tenths. · The loss by reflection is often one-half. In view of this peculiarity and many others, it is held that a twenty-six-inch refractor is fully equal to any six-foot reflector.

The mounting of large telescopes demands the highest engineering ability. The whole instrument, with its vast weight of a twenty-six-inch glass lens, with its accompanying tube and appurtenances, must be pointed as nicely as a rifle, and held as steadily as the axis of the globe. To give it the required steadiness, the foundation on which it is placed is sunk deep in the earth, far from rail or other roads, and no part of the observatory is allowed to touch this support. When a star is once found, the earth swiftly rotates the telescope away from it, and it passes out of the field. To avoid this, clock-work is so arranged that the great telescope follows the star by the hour, if required. It will take a star at its eastern rising, and hold it constantly in view while it climbs to the meridian and sinks in the west (Fig. 15). The reflector demands still more difficult engineering. That of Lord Rosse has a metallic mirror

Fig. 15.—Cambridge Equatorial.

weighing six tons, a tube fifty-six feet long, which, with its appurtenances, weighs seven tons more. It moves between two walls only 20° east and west. The new Paris reflector (Fig. 16) has a much wider range of movement.

The Spectroscope.

A spectrum is a collection of the colors which are dispersed by a prism from any given light. If it is sunlight, it is a solar spectrum; if the source of light is a

Fig. 16.—New Paris Reflector.

star, candle, glowing metal, or gas, it is the spectrum of
a star, candle, glowing metal, or gas. An instrument
to see these spectra is called a spectroscope. Considering the infinite variety of light, and its easy modifica-
tion and absorption, we should expect an immense
number of spectra. A mere prism disperses the light
so imperfectly that different orders of vibrations, perceived as colors, are mingled. No eye can tell where
one commences or ends. Such a spectrum is said to
be impure. What we want is that each point in the
spectrum should be made of rays of the same number
of vibrations. As we can let only a small beam of light
pass through the prism, in studying celestial objects with
a telescope and spectroscope we must, in every instance,

contract the aperture
of the instrument un-
til we get only a small
beam of light. In or-
der to have the colors
thoroughly dispersed,
the best instruments
pass the beam of light
through a series of
prisms called a bat-
tery, each one spread-
ing farther the colors
which the previous
ones had spread. In
Fig. 17 the ray is seen
entering through the

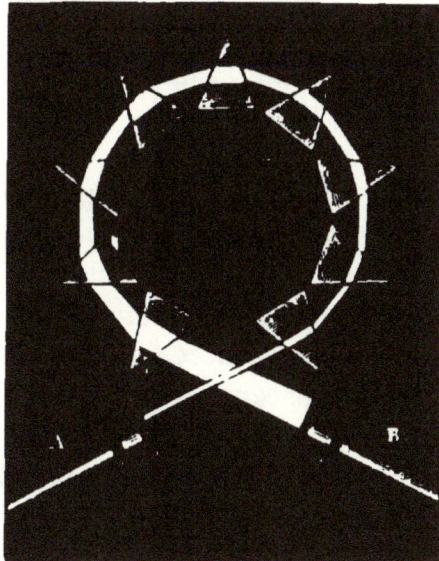

Fig. 17.—Spectroscope, with Battery of Prisms.

telescope A, which renders the rays parallel, and passing through the prisms out to telescope B, where the

3

spectrum can be examined on the retina of the eye for a screen. In order to still farther disperse the rays, some batteries receive the ray from the last prism at 0 upon an oblique mirror, send it up a little to another, which delivers it again to the prism to make its journey back again through them all, and come out to be examined just above where it entered the first prism.

Attached to the examining telescope is a diamond-ruled scale of glass, enabling us to fix the position of any line with great exactness.

In Fig. 18 is seen, in the lower part, a spectrum of the sun, with about a score of its thousands of lines

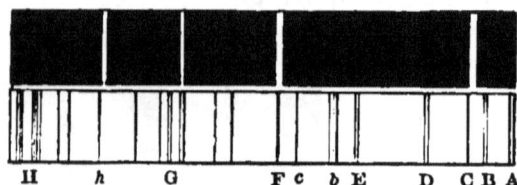

Fig. 18.—Spectra of glowing Hydrogen and the Sun.

made evident. In the upper part is seen the spectrum of bright lines given by glowing hydrogen gas. These lines are given by no other known gas; they are its autograph. It is readily observed that they precisely correspond with certain dark lines in the solar spectrum. Hence we easily know that a glowing gas gives the same bright lines that it absorbs from the light of another source passing through it — that is, glowing gas gives out the same rays of light that it absorbs when it is not glowing.

We are sure, therefore, that hydrogen exists in the sun. In the same manner we detect salt, iron, and a dozen other metals. Applied to the stars the spectro-

scope shows that they resemble the sun in constitution
and general condition. They are divided into four general orders, according to resemblances of their spectra.
The first order includes mostly stars showing a white
light, as Rigel, Altair, Regulus, etc. Nearly one-half of
the stars are included in this order. The second order
includes mostly stars showing a yellow light, as Arcturus, Aldebaran, etc. These most resemble the sun in
condition and chemical condition. The third order
shows a red light. The fourth includes only faint stars.

A patient study of these signs of substances reveals
richer results than a study of the cuniform characters
engraved on Assyrian slabs; for one is the handwriting of men, the other the handwriting of God.

One of the most difficult and delicate problems solved
by the spectroscope is the approach or departure of a
light-giving body in the line of sight. Stand before a
locomotive a mile away, you cannot tell whether it approaches or recedes, yet it will dash by in a minute.
How can the movements of the stars be comprehended
when they are at such an immeasurable distance?

It can best be illustrated by music. The note C of
the G clef is made by two hundred and fifty-seven vibrations of air per second. Twice as many vibrations
per second would give us the note C an octave above.
Sound travels at the rate of three hundred and sixty-
four yards per second. If the source of these two hundred and fifty-seven vibrations could approach us at
three hundred and sixty-four yards per second, it is obvious that twice as many waves would be put into a
given space, and we should hear the upper C when only
waves enough were made for the lower C. The same

result would appear if we carried our ear toward the
sound fast enough to take up twice as many waves as
though we stood still. This is apparent to every observer in a railway train. The whistle of an approaching locomotive gives one tone; it passes, and we instantly detect another. Let two trains, running at a
speed of thirty-six yards a second, approach each other. Let the whistle of one sound the note E, three
hundred and twenty-three vibrations per second. It
will be heard on the other as the note G, three hundred and eighty-eight vibrations per second; for the
speed of each train crowds the vibrations into one-tenth
less room, adding 32+ vibrations per second, making
three hundred and eighty-eight in all. The trains pass.
The vibrations are put into one-tenth more space by
the whistle making them, and the other train allows
only nine-tenths of what there are to overtake the ear.
Each subtracts 32+ vibrations from three hundred and
twenty-three, leaving only two hundred and fifty-eight,
which is the note C. Yet the note E was constantly
uttered.

If a source of light approach or depart, it will have a
similar effect on the light waves. How shall we detect
it? If a star approach us, it puts a greater number of
waves into an inch, and shortens their length. If it recedes, it increases the length of the wave—puts a less
number into an inch. If a body giving only the number of vibrations we call green were to approach sufficiently fast, it would crowd in vibrations enough to
appear what we call blue, indigo, or even violet, according to its speed. If it receded sufficiently fast, it would
leave behind it only vibrations enough to fill up the

space with what we call yellow, orange, or red, according to its speed; yet it would be green, and green only, all the time. But how detect the change? If red waves are shortened they become orange in color; and from below the red other rays, too far apart to be seen by the eye, being shortened, become visible as red, and we cannot know that anything has taken place. So, if a star recedes fast enough, violet vibrations being lengthened become indigo; and from above the violet other rays, too short to be seen, become lengthened into visible violet, and we can detect no movement of the colors. The dark lines of the spectrum are the cutting out of rays of definite wave-lengths. If the color spectrum moves away, they move with it, and away from their proper place in the ordinary spectrum. If, then, we find them toward the red end, the star is receding; if toward the violet end, it is approaching. Turn the instrument on the centre of the sun. The dark lines take their appropriate place, and are recognized on the ruled scale. Turn it on one edge, that is approaching us one and a quarter miles a second by the revolution of the sun on its axis, the spectral lines move toward the violet end; turn the spectroscope toward the other edge of the sun, it is receding from us one and a quarter miles a second by reason of the axial revolution, and the spectral lines move toward the red end. Turn it near the spots, and it reveals the mighty up-rush in one place and the down-rush in another of one hundred miles a second. We speak of it as an easy matter, but it is a problem of the greatest delicacy, almost defying the mind of man to read the movements of matter.

It should be recognized that Professor Young, of

Princeton, is the most successful operator in this recent realm of science.   He already proposes to correct the former estimate of the sun's axial rotations, derived from observing its spots, by the surer process of observing accelerated and retarded light.

Within a very few years this wonderful instrument, the spectroscope, has made amazing discoveries.   In chemistry it reveals substances never known before; in analysis it is delicate to the detection of the millionth of a grain.   It is the most deft handmaid of chemistry, the arts, of medical science, and astronomy. It tells the chemical constitution of the sun, the movements taking place, the nature of comets, and nebulæ. By the spectroscope we know that the atmospheres of Venus and Mars are like our own; that those of Jupiter and Saturn are very unlike; it tells us which stars approach and which recede, and just how one star differeth from another in glory and substance.

In the near future we shall have the brilliant and diversely colored flowers of the sky as well classified into orders and species as are the flowers of the earth.

IV.

CELESTIAL MEASUREMENTS.

"Who hath measured the waters in the hollow of his hand, and meted out heaven with the span? Mine hand also hath laid the foundation of the earth, and my right hand hath spanned the heavens."—*Isa.* xl. 12; xlviii. 13.

"Go to yon tower, where busy science plies
   Her vast antennæ, feeling thro' the skies;
   That little vernier, on whose slender lines
   The midnight taper trembles as it shines,
   A silent index, tracks the planets' march
   In all their wanderings thro' the ethereal arch,
   Tells through the mist where dazzled Mercury burns,
   And marks the spot where Uranus returns.

"So, till by wrong or negligence effaced,
   The living index which thy Maker traced
   Repeats the line each starry virtue draws
   Through the wide circuit of creation's laws:
   Still tracks unchanged the everlasting ray
   Where the dark shadows of temptation stray;
   But, once defaced, forgets the orbs of light,
   And leaves thee wandering o'er the expanse of night."

                              OLIVER WENDELL HOLMES.

IV.

CELESTIAL MEASUREMENTS.

WE know that astronomy has what are called practical uses. If a ship had been driven by Euroclydon for fourteen dismal days and nights without sun or star appearing, a moment's glance into the clear sky from the heaving deck, by a very slightly educated sailor, would tell within one hundred rods where he was, and determine the distance and way to the nearest port. We know that, in all final and exact surveying, positions must be fixed by the stars. Earth's landmarks are uncertain and easily removed; those which we get from the heavens are stable and exact.

In 1878 the United States steam-ship *Enterprise* was sent to survey the Amazon. Every night a "star party" went ashore to fix the exact latitude and longitude by observations of the stars. Our real landmarks are not the pillars we rear, but the stars millions of miles away. All our standards of time are taken from the stars; every railway train runs by their time to avoid collision; by them all factories start and stop. Indeed, we are ruled by the stars even more than the old astrologers imagined.

Man's finest mechanism, highest thought, and broadest exercise of the creative faculty have been inspired by astronomy. No other instruments approximate in delicacy those which explore the heavens; no other

3*

system of thought can draw such vast and certain conclusions from its premises. " Too low they build who build beneath the stars;" we should lay our foundations in the skies, and then build upward.

We have been placed on the outside of this earth, instead of the inside, in order that we may look abroad. We are carried about, through unappreciable distance, at the inconceivable velocity of one thousand miles a minute, to give us different points of vision. The earth, on its softly-spinning axle, never jars enough to unnest a bird or wake a child; hence the foundations of our observatories are firm, and our measurements exact. Whoever studies astronomy, under proper guidance and in the right spirit, grows in thought and feeling, and becomes more appreciative of the Creator.

Celestial Movements.

Let it not be supposed that a mastery of mathematics and a finished education are necessary to understand the results of astronomical research. It took at first the highest power of mind to make the discoveries that are now laid at the feet of the lowliest. It took sublime faith, courage, and the results of ages of experience in navigation, to enable Columbus to discover that path to the New World which now any little boat can follow. Ages of experience and genius are stored up in a locomotive, but quite an unlettered man can drive it. It is the work of genius to render difficult matters plain, abstruse thoughts clear.

A brief explanation of a few terms will make the principles of world inspection easily understood. Imagine a perfect circle thirty feet in diameter—that is, create one (Fig. 19). Draw through it a diameter horizontally, another perpendicularly. The angles made by the intersecting lines are each said to be ninety degrees, marked thus °. The arc of a circle included between any two of the lines is also 90°. Every circle, great or small, is divided into these 360°. If the sun rose in the east and came to the zenith at noon, it would have passed 90°. When it set in the west it would have traversed half the circle, or 180°. In Fig. 20 the angle of the lines meas-

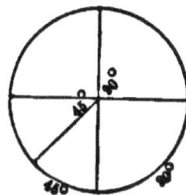

Fig. 19.

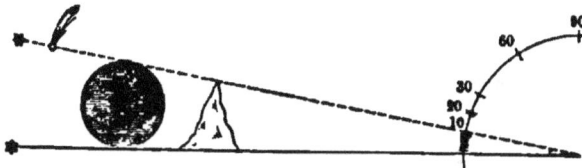

Fig. 20.—Illustration of Angles.

ured on the graduated arc is 10°. The mountain is 10° high, the world 10° in diameter, the comet moves 10° a day, the stars are 10° apart. The height of the mountain, the diameter of the world, the velocity of the comet, and the distance between the stars, depend on the distance of each from the point of sight. Every degree is divided into 60 minutes (marked ′), and every minute into 60 seconds (marked ″).

Imagine yourself inside a perfect sphere one hundred feet in diameter, with the interior surface above, around, and below studded with fixed bright points like stars. The familiar constellations of night might be blazoned there in due proportion.

If this star-sprent sphere were made to revolve once in twenty-four hours, all the stars would successively pass in review. How easily we could measure distances between stars, from a certain fixed meridian, or the equator! How easily we could tell when any particular star would culminate! It is as easy to take all these measurements when our earthly observatory is steadily revolved within the sphere of circumambient stars. Stars can be mapped as readily as the streets of a great city. Looking down on it in the night, one could trace the lines of lighted streets, and judge something of its extent and regularity. But the few lamps of evening would suggest little of the greatness of the public buildings, the magnificent enterprise and commerce of its citizens, or the intelligence of its scholars. Looking up to the lamps of the celestial city, one can judge something of its extent and regularity; but they suggest little of the magnificence of the many mansions.

Stars are reckoned as so many degrees, minutes, and seconds from each other, from the zenith, or from a given meridian, or from the equator. Thus the stars called the Pointers, in the Great Bear, are 5° apart; the nearest one is 29° from the Pole Star, which is some 40° above the horizon at Philadelphia. In going to England you creep up toward the north end of the earth, till the Pole Star is 54° high. It stays near its place among the stars continually,

> "Of whose true-fixed and resting quality
> There is no fellow in the firmament."

How to Measure.

Suppose a telescope, fixed to a mural circle, to revolve on an axis, as in Fig. 21; point it horizontally at a star;

turn it up perpendicular to another star. Of course the two stars are 90° apart, and the graduated scale, which is attached to the outer edge of the circle, shows a revolution of a quarter circle, or 90°. But a perfect accuracy of measurement must be sought; for a mistake of the breadth of a hair, seen at the distance of one hundred and twenty-five feet, would cause an error of 3,000,000 miles at the distance of the sun, and im-

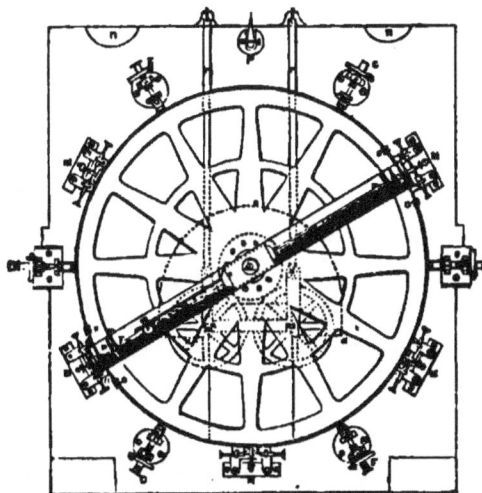

Fig. 21.—Mural Circle.

mensely more at the distance of the stars. The correction of an inaccuracy of no greater magnitude than that has reduced our estimate of the distance of our sun 3,000,000 miles.

Consider the nicety of the work. Suppose the graduated scale to be thirty feet in circumference. Divided into 360°, each would be one inch long. Divide each degree into 60', each one is $\frac{1}{60}$ of an inch long. It takes good eyesight to discern it. But each minute must be divided into 60″, and these must not only be noted, but even tenths and hundredths of seconds must be discerned. Of course they are not seen by the naked eye; some mechanical contrivance must be called in to assist. A watch loses two minutes a week, and hence is unreliable. It is taken to a watch-maker that every single second may be quickened $\frac{1}{20160}$ part of itself. Now $\frac{1}{30000}$ part of a second would be a small interval of time to measure, but it must be under control. If the temperature of a summer morning rises ten or twenty degrees we scarcely notice it; but the magnetic tasimeter measures $\frac{1}{5000}$ of a degree.

Come to earthly matters. In 1874, after nearly twenty-eight years' work, the State of Massachusetts opened a tunnel nearly five miles long through the Hoosac Mountains. In the early part of the work the engineers sunk a shaft near the middle 1028 feet deep. Then the question to be settled was where to go so as to meet the approaching excavations from the east and west. A compass could not be relied on under a mountain. The line must be mechanically fixed. A little divergence at the starting-point would become so great, miles away, that the excavations might pass each other without meeting; the grade must also rise toward the central shaft, and fall in working away from it; but the lines were fixed with such infinitesimal accuracy that, when the one going west from the eastern portal and the one going east from the shaft met in the heart of the mountain, the western line was only one-eighth of an inch too high, and three-sixteenths of an inch too far north. To reach this perfect result they had to triangulate from the eastern portal to distant mountain peaks, and thence down the valley to the central shaft, and thus fix the direction of the proposed line across the mouth of the shaft. Plumb-lines were then dropped one thousand and twenty-eight feet, and thus the line at the bottom was fixed.

Three attempts were made—in 1867, 1870, and 1872 —to fix the exact time-distance between Greenwich and Washington. These three separate efforts do not differ one-tenth of a second. Such demonstrable results on earth greatly increase our confidence in similar measurements in the skies.

A scale is frequently affixed to a pocket-rule, by which we can easily measure one-hundredth of an inch (Fig. 22). The upper and lower line is divided into tenths of an inch. Observe the slanting line at the right hand. It leans from the perpendicular one-tenth of an inch, as shown by noticing where it reaches the top line. When it reaches the second horizontal line it

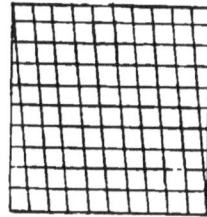

Fig. 22.

has left the perpendicular one-tenth of that tenth—that is, one-hundredth. The intersection marks $\frac{99}{100}$ of an inch from one end, and one-hundredth from the other.

When division-lines, on measures of great nicety, get too fine to be read by the eye, we use the microscope. By its means we are able to count 112,000 lines ruled on a glass plate within an inch. The smallest object that can be seen by a keen eye makes an angle of $40''$, but by putting six microscopes on the scale of the telescope on the mural circle, we are able to reach an exactness of $0''.1$, or $\frac{1}{3600}$ of an inch. This instrument is used to measure the declination of stars, or angular distance north or south of the equator. Thus a star's place in two directions is exactly fixed. When the telescope is mounted on two pillars instead of the face of a wall, it is called a transit instrument. This is used to determine the time of transit of a star over the meridian, and if the transit instrument is provided with a graduated circle it can also be used for the same purposes as the mural circle. Man's capacity to measure exactly is indicated in his ascertainment of the length of waves of light. It is easy to measure the three hundred feet distance between the crests of storm-waves in the wide Atlantic; easy to measure the different wave-lengths of the different tones of musical sounds. So men measure the lengths of the undulations of light. The shortest is of the violet light, 154.84 ten-millionths of an inch. By the horizontal pendulum Professor Root has made $\frac{1}{36000000}$ of an inch apparent.

The next elements of accuracy must be perfect time and perfect notation of time. As has been said, we get our time from the stars. Thus the infinite and heavenly dominates the finite and earthly. Clocks are set to the invariable sidereal time. Sidereal noon is when we have turned ourselves under the point where the sun crosses the equator in March, called the vernal equinox. Sidereal clocks are figured to indicate twenty-four hours in a day: they tick exact seconds. To map stars we wish to know the exact second when they cross the meridian, or the north and south line in the celestial dome above us. The telescope (Fig. 21, p. 61) swings exactly north and south. In its focus a set of fiue threads of spider-lines is placed (Fig. 23). The telescope is set just high enough, so that by the rolling over of the earth

the star will come into the field just above the horizontal thread. The observer
notes the exact second and
tenth of a second when the
star reaches each vertical
thread in the instrument,
adds together the times and
divides by five to get the
average, and the exact time
is reached.

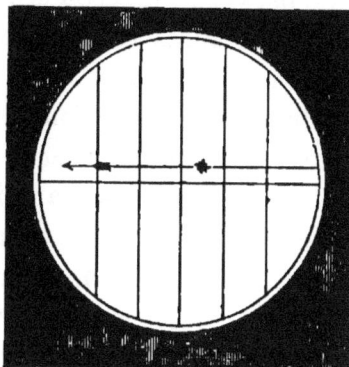

Fig. 23.—Transit of a Star noted.

But man is not reliable
enough to observe and record
with sufficient accuracy. Some, in their excitement, an-
ticipate its positive passage, and some cannot get their
slow mental machinery in motion till after it has made
the transit. Moreover, men fall into a habit of estimating some numbers of tenths of a second oftener
than others. It will be found that a given observer
will say three tenths or seven tenths oftener than four
or eight. He is falling into ruts, and not trustworthy.
General O. M. Mitchel, who had been director of the
Cincinnati Observatory, once told one of his staff - offi-
cers that he was late at an appointment. "Only a
few minutes," said the officer, apologetically. "Sir,"
said the general, "where I have been accustomed to
work, hundredths of a second are too important to be
neglected." And it is to the rare genius of this astronomer, and to others, that we owe the mechanical accuracy that we now attain. The clock is made to mark its
seconds on paper wrapped around a revolving cylinder.
Under the observer's fingers is an electric key. This
he can touch at the instant of the transit of the star

over each wire, and thus put his observation on the same line between the seconds dotted by the clock. Of course these distances can be measured to minute fractional parts of a second.

But it has been found that it takes an appreciable time for every observer to get a thing into his head and out of his finger-ends, and it takes some observers longer than others. A dozen men, seeing an electric spark, are liable to bring down their recording marks in a dozen different places on the revolving paper. Hence the time that it takes for each man to get a thing into his head and out of his fingers is ascertained. This time is called his personal equation, and is subtracted from all of his observations in order to get at the true time; so willing are men to be exact about material matters. Can it be thought that moral and spiritual matters have no precision? Thus distances east or west from any given star or meridian are secured; those north and south from the equator or the zenith are as easily fixed, and thus we make such accurate maps of the heavens that any movements in the far-off stars—so far that it may take centuries to render the swiftest movements appreciable—may at length be recognized and account-ed for.

We now come to a little study of the modes of measuring distances. Create a perfect square (Fig. 24); draw a diagonal line. The square angles are 90°, the divided angles give two of 45° each. Now the base A B is equal to the perpendicu-lar A C. Now any point—C, where a

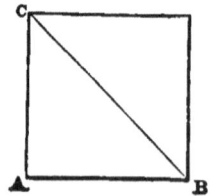

Fig. 24.

perpendicular, A C, and a diagonal, B C, meet—will be

as far from A as B is. It makes no difference if a river
flows between A and C, and we cannot go over it; we
can measure its distance as easily as
if we could. Set a table four feet by
eight out-doors (Fig. 25); so arrange it
that, looking along one end, the line
of sight just strikes a tree the other
side of the river. Go to the other
end, and, looking toward the tree, you
find the line of sight to the tree falls
an inch from the end of the table on
the farther side. The lines, therefore,
approach each other one inch in every
four feet, and will come together at a
tree three hundred and eighty-four
feet away.

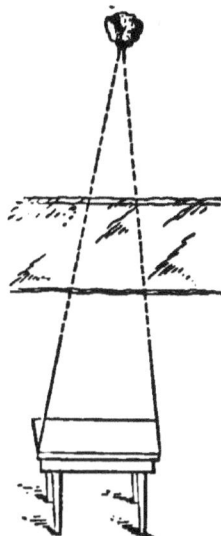

Fig. 25.—Measuring Distances.

The next process is to measure the
height or magnitude of objects at an
ascertained distance. Put two pins in a stick half an
inch apart (Fig. 26). Hold it up two feet from the eye,

Fig. 26.—Measuring Elevations.

and let the upper pin fall in line with your eye and the
top of a distant church steeple, and the lower pin in
line with the bottom of the church and your eye. If
the church is three-fourths of a mile away, it must be
eighty-two feet high; if a mile away, it must be one
hundred and ten feet high. For if two lines spread

one-half an inch going two feet, in going four feet they will spread an inch, and in going a mile, or five thousand two hundred and eighty feet, they will spread out one-fourth as many inches, viz., thirteen hundred and twenty—that is, one hundred and ten feet. Of course these are not exact methods of measurement, and would not be correct to a hair at one hundred and twenty-five feet, but they perfectly illustrate the true methods of measurement.

Imagine a base line ten inches long. At each end erect a perpendicular line. If they are carried to infinity they will never meet: will be forever ten inches apart. But at the distance of a foot from the base line incline one line toward the other $\frac{63}{10000000}$ of an inch, and the lines will come together at a distance of three hundred miles. That new angle differs from the former right angle almost infinitesimally, but it may be measured. Its value is about three-tenths of a second. If we lengthen the base line from ten inches to all the miles we can command, of course the point of meeting will be proportionally more distant. The angle made by the lines where they come together will be obviously the same as the angle of divergence from a right angle at this end. That angle is called the parallax of any body, and is the angle that would be made by two lines coming from that body to the two ends of any conventional base, as the semi-diameter of the earth. That that angle would vary according to the various distances is easily seen by Fig. 27.

Let O P be the base. This would subtend a greater angle seen from star A than from star B. Let B be far enough away, and O P would become invisible, and B would have no parallax for that base. Thus the moon has a parallax of 57′ with the semi-equatorial diameter of the earth for a base. And the sun has a parallax 8″.85 on the same base. It is not necessary to confine ourselves to right angles in these measurements, for the same principles hold true in any angles. Now, suppose two observers on the equator should look at the moon at the same instant. One is on the top of Cotopaxi, on the west coast of South America, and one on the west coast of Africa. They are 90° apart—half the earth's diameter between them. The one on Cotopaxi sees it exactly overhead, at an angle of 90° with the earth's diameter. The one on the coast of Africa sees its angle with the same line to be 89° 3′—that is, its parallax is 57′. Try the same

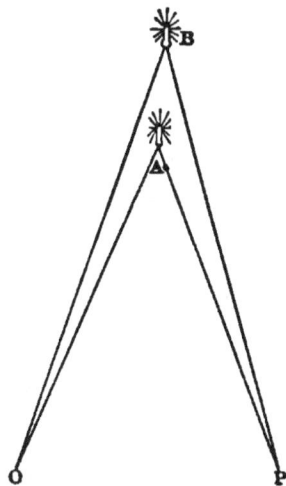

Fig. 27.

experiment on the sun farther away, as is seen in Fig. 27, and its smaller parallax is found to be only 8″.85.

It is not necessary for two observers to actually station themselves at two distant parts of the earth in order to determine a parallax. If an observer could go from one end of the base-line to the other, he could determine both angles. Every observer is actually carried along through space by two motions: one is that of the earth's revolution of one thousand miles an hour around the axis; and the other is the movement of the earth around the sun of one thousand miles in a minute. Hence we can have the diameter not only of the earth (eight thousand miles) for a base-line, but the diameter of the earth's orbit (185,000,000 miles), or any part of it, for such a base. Two observers at the ends of the earth's diameter, looking at a star at the same instant, would find that it made the same angle at both ends; it has no parallax on so short a base. We must seek a longer one. Observe a certain star on the 21st of March; then let us traverse the realms of space for six months, at one thousand miles a minute. We come round in our orbit to a point opposite where we were six months ago, with 184,000,000 of miles between the points. Now, with this for a base-line, measure the angles of the same stars: it is the same angle. Sitting in my study here, I glance out of the window and discern separate bricks, in houses five hundred feet away, with my unaided eye; they subtend a discernible angle. But one thousand feet away I cannot distinguish individual bricks; their width, being only two inches, does not subtend an angle apprehensible to my vision. So at these distant stars the earth's enormous orbit, if lying like a blazing ring in space, with the world set on its edge like a pearl, and the sun blazing like a diamond in the centre, would all shrink to a mere point. Not quite to a point from the nearest stars, or we should never be able to measure the distance of any of them. Professor Airy says that our orbit, seen from the nearest star, would be the same as a circle six-tenths of an inch in diameter seen at the distance of a mile: it would all be hidden by a thread one-twenty-fifth of an inch in diameter, held six hundred and fifty feet from the eye. If a straight line could be drawn from a star, Sirius in the east to the star Vega in the west, touching our

earth's orbit on one side, as T R A (Fig. 28), and a line
were to be drawn six
months later from the
same stars, touching our
earth's orbit on the oth-
er side, as R B T, such a
line would not diverge sufficiently from a straight line
for us to detect its divergence. Numerous vain attempts had been made, up to the year 1835, to detect
and measure the angle of parallax by which we could
rescue some one or more of the stars from the inconceivable depths of space, and ascertain their distance from
us. We are ever impelled to triumph over what is declared to be unconquerable. There are peaks in the
Alps no man has ever climbed. They are assaulted
every year by men zealous of more worlds to conquer.
So these greater heights of the heavens have been assaulted, till some ambitious spirits have outsoared even
imagination by the certainties of mathematics.

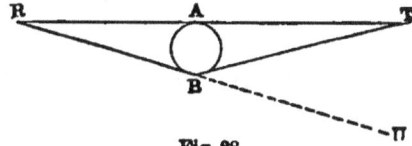

It is obvious that if one star were three times as far
from us as another, the nearer one would seem to be
displaced by our movement in our orbit three times as
much as the other; so, by comparing one star with another, we reach a ground of judgment. The ascertainment of longitude at sea by means of the moon affords
a good illustration. Along the track where the moon
sails, nine bright stars, four planets, and the sun have
been selected. The nautical almanacs give the distance
of the moon from these successive stars every hour in the
night for three years in advance. The sailor can measure
the distance at any time by his sextant. Looking from
the world at D (Fig. 29), the distance of the moon and

star is A E, which is given in the almanac.   Looking
from C, the distance is only B E, which enables even the
uneducated sailor to find the distance, C D, on the earth,
or his distance from Greenwich.

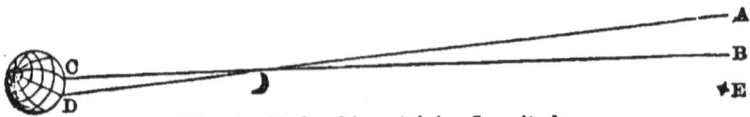

Fig. 29.—Mode of Ascertaining Longitude.

So, by comparisons of the near and far stars, the approximate distance of a few of them has been determined.   The nearest one is the brightest star in the
Centaur, seldom visible in our northern latitudes, which
has a parallax of about one second.   The next nearest is
No. 61 in the Swan, or 61 Cygni, having a parallax of
$0''.34$.   Approximate measurements have been made on
Sirius, Capella, the Pole Star, etc., about eighteen in all.
The distances are immense: only the swiftest agents
can traverse them.   If our earth were suddenly to dis-
solve its allegiance to the king of day, and attempt a
flight to the North Star, and should maintain its flight
of one thousand miles a minute, it would fly away toward Polaris for thousands upon thousands of years, till
a million years had passed away, before it reached that
northern dome of the distant sky, and gave its new allegiance to another sun.   The sun it had left behind it
would gradually diminish till it was small as Arcturus,
then small as could be discerned by the naked eye,
until at last it would finally fade out in utter darkness
long before the new sun was reached.   Light can traverse the distance around our earth eight times in one
second.   It comes in eight minutes from the sun, but
it takes three and a quarter years to come from Alpha

Centauri, seven and a quarter years from 61 Cygni, and forty-five years from the Polar Star.

Sometimes it happens that men steer along a lee shore, dependent for direction on Polaris, that light-house in the sky. Sometimes it has happened that men have traversed great swamps by night when that star was the light-house of freedom. In either case the exigency of life and liberty was provided for forty-five years before by a Providence that is divine.

We do not attempt to name in miles these enormous distances; we must seek another yard-stick. Our astronomical unit and standard of measurement is the distance of the earth from the sun — 92,500,000 miles. This is the golden reed with which we measure the celestial city. Thus, by laying down our astronomical unit 226,000 times, we measure to Alpha Centauri, more than twenty millions of millions of miles. Doubtless other suns are as far from Alpha Centauri and each other as that is from ours.

Stars are not near or far according to their brightness. 61 Cygni is a telescopic star, while Sirius, the brightest star in the heavens, is twice as far away from us. One star differs from another star in intrinsic glory.

The highest testimonies to the accuracy of these celestial observations are found in the perfect predictions of eclipses, transits of planets over the sun, occultation of stars by the moon, and those statements of the Nautical Almanac that enable the sailor to know exactly where he is on the pathless ocean by the telling of the stars: "On the trackless ocean this book is the mariner's trusted friend and counsellor; daily and nightly its revelations bring safety to ships in all parts of the

4

world. It is something more than a mere book; it is an ever-present manifestation of the order and harmony of the universe."

Another example of this wonderful accuracy is found in tracing the asteroids. Within 200,000,000 or 300,000,000 miles from the sun, the two hundred and fifty (September, 1885,) minute bodies that have been already discovered move in paths very nearly the same—indeed two of them traverse the same orbit, being one hundred and eighty degrees apart;—they look alike, yet the eye of man in a few observations so determines the curve of each orbit, that one is never mistaken for another. But astronomy has higher uses than fixing time, establishing landmarks, and guiding the sailor. It greatly quickens and enlarges thought, excites a desire to know, leads to the utmost exactness, and ministers to adoration and love of the Maker of the innumerable suns.

V.

THE SUN.

"And God made two great lights; the greater light to rule the day, and the lesser light to rule the night: he made the stars also."—*Gen.* i. 16.

"It is perceived that the sun of the world, with all its essence, which is heat and light, flows into every tree, and into every shrub and flower, and into every stone, mean as well as precious; and that every object takes its portion from this common influx, and that the sun does not divide its light and heat, and dispense a part to this and a part to that. It is similar with the sun of heaven, from which the Divine love proceeds as heat, and the Divine wisdom as light; these two flow into human minds, as the heat and light of the sun of the world into bodies, and vivify them according to the quality of the minds, each of which takes from the common influx as much as is necessary."—SWEDENBORG.

V.

THE SUN.

Suppose we had stood on the dome of Boston State-house November 9th, 1872, on the night of the great conflagration, and seen the fire break out; seen the engines dash through the streets, tracking their path by their sparks; seen the fire encompass a whole block, leap the streets on every side, surge like the billows of a storm-swept sea; seen great masses of inflammable gas rise like dark clouds from an explosion, then take fire in the air, and, cut off from the fire below, float like argosies of flame in space. Suppose we had felt the wind that came surging from all points of the compass to fan that conflagration till it was light enough a mile away to see to read the finest print, hot enough to decompose the torrents of water that were dashed on it, making new fuel to feed the flame. Suppose we had seen this spreading fire seize on the whole city, extend to its environs, and, feeding itself on the very soil, lick up Worcester with its tongues of flame—Albany, New York, Chicago, St. Louis, Cincinnati—and crossing the plains swifter than a prairie fire, making each peak of the Rocky Mountains hold up aloft a separate torch of flame, and the Sierras whiter with heat than they ever were with snow, the waters of the Pacific resolve into their constituent elements of oxygen and hydrogen, and

burn with unquenchable fire! We withdraw into the
air, and see below a world on fire. All the prisoned
powers have burst into intensest activity. Quiet breezes
have become furious tempests. Look around this flaming globe—on fire above, below, around—there is nothing but fire. Let it roll beneath us till Boston comes
round again. No ember has yet cooled, no spire of
flame has shortened, no surging cloud has been quiet-
ed. Not only are the mountains still in flame, but
other ranges burst up out of the seething sea. There is
no place of rest, no place not tossing with raging flame!
Yet all this is only a feeble figure of the great burning sun. It is but the merest hint, a million times too
insignificant.

The sun appears small and quiet to us because we are
so far away. Seen from the various planets, the relative size of the sun appears as in Fig. 30. Looked for
from some of the stars about us, the sun could not be
seen at all. Indeed, seen from the earth, it is not always the same size, because the distance is not always
the same. If we represent the size of the sun by one
thousand on the 23d of September or 21st of March, it
would be represented by nine hundred and sixty-seven
on the 1st of July, and by one thousand and thirty-four
on the 1st of January.

We sometimes speak of the sun as having a diameter
of 860,000 miles. We mean that that is the extent of
the body as seen by the eye. But that is a small part
of its real diameter. So we say the earth has an equatorial diameter of 7925$\frac{1}{4}$ miles, and a polar one of 7899.
But the air is as much a part of the earth as the rocks
are. The electric currents are as much a part of the

earth as the ores and mountains they traverse. What the diameter of the earth is, including these, no man can tell. We used to say the air extended forty-five miles, but we now know that it reaches vastly farther. So of

Fig. 30.—Relative Size of Sun as seen from Different Planets.

the sun, we might almost say that its diameter is infinite, for its light and heat reach beyond our measurement. Its living, throbbing heart sends out pulsations, keeping all space full of its tides of living light.

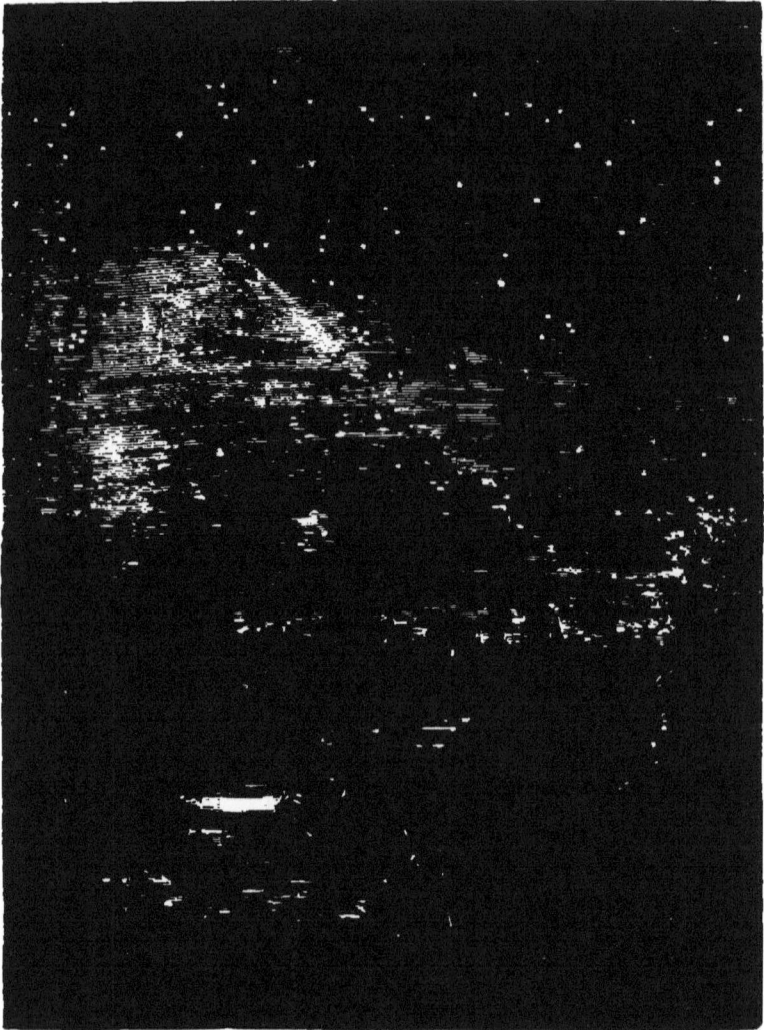

Fig. 31.—Zodiacal Light.

We might say with evident truth that the far-off planets are a part of the sun, since the space they traverse is filled with the power of that controlling king; not only with light, but also with gravitating power.

But come to more ponderable matters. If we look

into our western sky soon after sunset, on a clear, moon-less night in March or April, we shall see a dim, soft light, somewhat like the milky-way, often reaching, well defined, to the Pleiades. It is wedge-shaped, in-clined to the south, and the smallest star can easily be seen through it. Mairan and Cassini affirm that they have seen sudden sparkles and movements of light in it. All our best tests show the spectrum of this light to be continuous, and therefore reflected; which indicates that it is a ring of small masses of meteoric matter surround-ing the sun, revolving with it and reflecting its light. One bit of stone as large as the end of one's thumb, in a cubic mile, would be enough to reflect what light we see looking through millions of miles of it. Perhaps an eye sufficiently keen and far away would see the sun surrounded by a luminous disk, as Saturn is with his rings. As it extends beyond the earth's orbit, if this be measured as a part of the sun, its diameter would be about 200,000,000 miles.

Come closer. When the sun is covered by the disk of the moon at the instant of total eclipse, observers are startled by strange swaying luminous banners, ghostly and weird, shooting in changeful play about the central darkness (Fig. 32). These form the corona. Men have usually been too much moved to describe them, and have always been incapable of drawing them in the short minute or two of their continuance. But in 1878 men travelled eight thousand miles, coming and return-ing, in order that they might note the three minutes of total eclipse in Colorado. Each man had his work as-signed to him, and he was drilled to attend to that and nothing else. Improved instruments were put into his

4*

hands, so that the sun was made to do his own drawing
and give his own picture at consecutive instants.   Fig.
33 is a copy of a photograph of the corona of 1878, by

Fig. 32.—The Corona in 1858, Brazil.

Mr. Henry Draper.   It showed much less changeability
that year than common, it being very near the time of
least sun-spot. The previous picture was taken near the
time of maximum sun-spot.

It was then settled that the corona consists of reflected light, sent to us from dust particles or meteoroids swirling in the vast seas, giving new densities and

rarities, and hence this changeful light. Whether they are there by constant projection, and fall again to the sun, or are held by electric influence, or by force of orbital revolution, we do not know. That the corona cannot be in any sense an atmosphere of any continuous gas, is seen from the fact that the comet of 1843, passing within 93,000 miles of the body of the sun, was not burned out of existence as a comet, nor in any percepti-

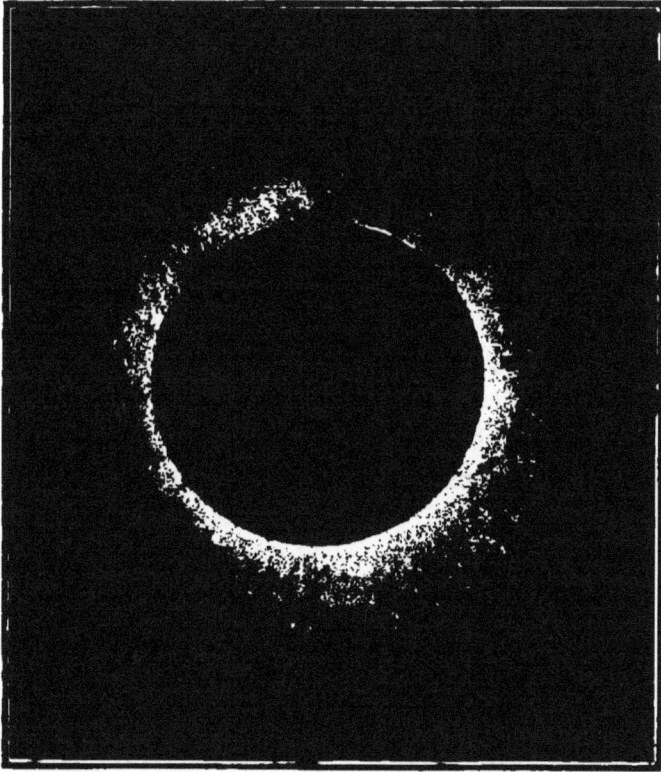

Fig. 33.—The Corona in 1878, Colorado.

ble degree retarded in its motion. If the sun's diameter is to include the corona, it will be from 1,260,000 to 1,460,000 miles.

Come closer still. At the instant of the totality of the eclipse red flames of most fantastic shape play along the edge of the moon's disk. They can be seen at any time by the use of a proper telescope with a spectroscope attached. I have seen them with great distinctness and brilliancy with the excellent eleven-inch telescope of the Wesleyan University. A description of their appearance is best given in the language of Professor Young, of Princeton College, who has made these flames the object of most successful study. On September 7th, 1871, he was observing a large hydrogen cloud by the sun's edge. This cloud was about 100,000 miles long, and its upper side was some 50,000 miles above the sun's surface, the lower side some 15,000 miles. The whole had the appearance of being supported on pillars of fire, these seeming pillars being in reality hydrogen jets brighter and more active than the substance of the cloud. At half-past twelve, when Professor Young chanced to be called away from his observatory, there were no indications of any approaching change, except that one of the connecting stems of the southern extremity of the cloud had grown considerably brighter and more curiously bent to one side; and near the base of another, at the northern end, a little brilliant lump had developed itself, shaped much like a summer thunderhead.

But when Professor Young returned, about half an hour later, he found that a very wonderful change had taken place, and that a very remarkable process was actually in progress. "The whole thing had been literally blown to shreds," he says, "by some inconceivable uprush from beneath. In place of the quiet cloud I had

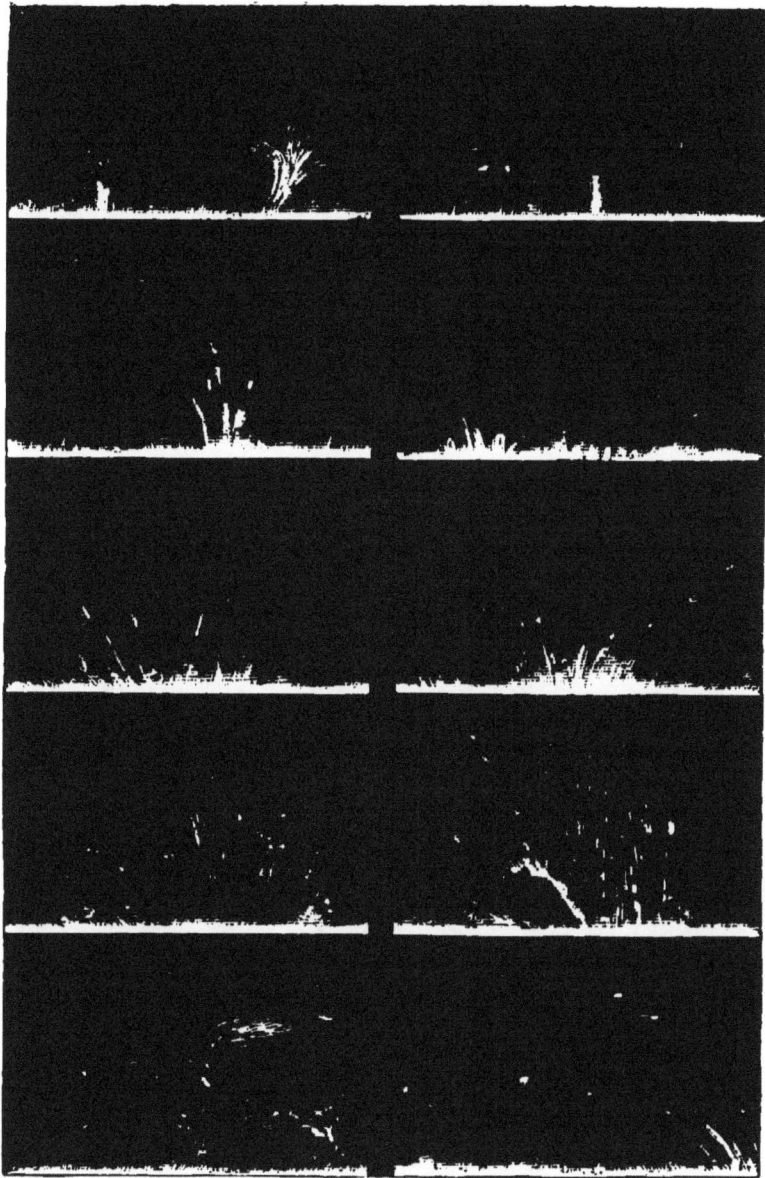

Fig. 34.—Solar Prominences of Flaming Hydrogen.

left, the air—if I may use the expression—was filled
with the flying *débris*, a mass of detached vertical fusi-
form fragments, each from ten to thirty seconds (*i. e.*,
from four thousand five hundred to thirteen thousand
five hundred miles) long, by two or three seconds (nine
hundred to thirteen hundred and fifty miles) wide—
brighter, and closer together where the pillars had formerly stood, and rapidly ascending. When I looked,
some of them had already reached a height of nearly
four minutes (100,000 miles); and while I watched
them they arose with a motion almost perceptible to
the eye, until, in ten minutes, the uppermost were more
than 200,000 miles above the solar surface. This was
ascertained by careful measurements, the mean of three
closely accordant determinations giving 210,000 miles
as the extreme altitude attained. I am particular in
the statement, because, so far as I know, chromato-
spheric matter (red hydrogen in this case) has never
before been observed at any altitude exceeding five
minutes, or 135,000 miles. The velocity of ascent, also
—one hundred and sixty-seven miles per second—is
considerably greater than anything hitherto recorded.
* * * As the filaments arose, they gradually faded away
like a dissolving cloud, and at a quarter past one only
a few filmy wisps, with some brighter streamers low
down near the chromatosphere, remained to mark the
place. But in the mean while the little 'thunder-head'
before alluded to had grown and developed wonderfully into a mass of rolling and ever-changing flame,
to speak according to appearances. First, it was crowd-
ed down, as it were, along the solar surface; later, it
arose almost pyramidally 50,000 miles in height; then

its summit was drawn down into long filaments and threads, which were most curiously rolled backward and forward, like the volutes of an Ionic capital, and finally faded away, and by half-past two had vanished like the other. The whole phenomenon suggested most forcibly the idea of an explosion under the great prominence, acting mainly upward, but also in all directions outward; and then, after an interval, followed by a corresponding in-rush."

No language can convey nor mind conceive an idea of the fierce commotion we here contemplate. If we call these movements hurricanes, we must remember that what we use as a figure moves but one hundred miles an hour, while these move one hundred miles a second. Such storms of fire on earth, "coming down upon us from the north, would, in thirty seconds after they had crossed the St. Lawrence, be in the Gulf of Mexico, carrying with them the whole surface of the continent in a mass not simply of ruins but of glowing vapor, in which the vapors arising from the dissolution of the materials composing the cities of Boston, New York, and Chicago would be mixed in a single indistinguishable cloud." In the presence of these evident visions of an actual body in furious flame, we need hesitate no longer in accepting as true the words of St. Peter of the time "in which the [atmospheric] heavens shall pass away with a great noise, and the elements shall melt with fervent heat; the earth also, and the works that are therein, shall be burned up."

This region of discontinuous flame below the corona is called the chromosphere. Hydrogen is the principal material of its upper part; iron, magnesium, and other metals, some of them as yet unknown on earth, but having a record in the spectrum, in the denser parts below. If these fierce fires are a part of the sun, as they assuredly are, its diameter would be from 1,060,000 to 1,260,000 miles.

Let us approach even nearer. We see a clearly recognized even disk, of equal dimensions in every direction. This is the photosphere. We here reach some definitely measurable data for estimating its visible size. We already know its distance. Its disk subtends an angle of 32' 4" ± 2", or a little more than half a degree. Three hundred and forty such suns, laid side by side, would span the celestial arch from east to west with a half circle of light. Two lines drawn from our earth at the angle mentioned would be 860,000 miles apart at the distance of 92,500,000 miles. This, then, is the diameter of the visible and measurable part of the sun. It would require one hundred and eight globes like the earth in a line to measure the sun's diameter, and three hundred and thirty-nine, to be strung like the beads of a neck lace, to encircle his waist. The sun has a volume equal to 1,245,000 earths, but being about a quarter as dense, it has a mass of only 326,800 earths. It has seven hundred times the mass of all the planets, asteroids, and satellites put together. Thus it is able to control them all by its greater power of attraction.

Concerning the condition of the surface of the sun many opinions are held. That it is hot beyond all estimate is indubitable. Whether solid or gaseous we are not sure. Opinions differ: some incline to the first theory, others to the second; some deem the sun composed of solid particles, floating in gas so condensed

by pressure and attraction as to shine like a solid. It
has no sensible changes of general level, but has pro-
digious activity in spots. These spots have been the
objects of earnest and almost hourly study on the part
of such men as Secchi, Lockyer, Faye, Young, and others, for years. But it is a long way off to study an object. No telescope brings it nearer than 100,000 miles.
Theory after theory has been advanced, each one satisfactory in some points, none in all. The facts about the
spots are these: They are most abundant on the two
sides of the equator. They are gregarious, depressed
below the surface, of vast extent, black in the centre,
usually surrounded by a region of partial darkness, beyond which is excessive light. They have motion of
their own over the surface—motion rotating about an
axis, upward and downward about the edges. They
change their apparent shape as the sun carries them

Fig. 35.—Change in Spots as rotated across the Disk, showing Cavities.

across its disk by axial revolution, being narrow as they
present their edges to us, and rounder as we look perpendicularly into them (Fig. 35).

These spots are also very variable in number, sometimes there being none for nearly two hundred days, and
again whole years during which the sun is never without them. The period from maximum to maximum

of spots is about eleven years. We might have looked for them in vain in 1878. They were numerous in 1884, and will be in 1894 if not again delayed. The cause of this periodicity was inferred to be the near approach of the enormous planet Jupiter, causing disturbance by its attraction. But the periods do not correspond, and the cause is the result of some law of solar action to us as yet unknown.

These spots may be seen with almost any telescope, the eye being protected by deeply colored glasses.

Until within one hundred years they were supposed to be islands of scoriæ floating in the sea of molten matter. But they were depressed below the surface, and showed a notch when on the edge. Wilson originated and Herschel developed the theory that the sun's real body was dark, cool, and habitable, and that the photosphere was a luminous stratum at a distance from the real body, with openings showing the dark spots below. Such a sun would have cooled off in a week, but would previously have annihilated all life below.

The solar spots being most abundant on the two sides of the equator, indicates their cyclonic character; the centre of a cyclone is rarefied, and therefore colder, and cold on the sun is darkness. M. Faye says: " Like our cyclones, they are descending, as I have proved by a special study of these terrestrial phenomena. They carry down into the depths of the solar mass the cooler materials of the upper layers, formed principally of hydrogen, and thus produce in their centre a decided extinction of light and heat as long as the gyratory movement continues. Finally, the hydrogen set free at the base of the whirlpool becomes reheated at this

great depth, and rises up tumultuously around the whirlpool, forming irregular jets, which appear above the
chromosphere. These jets constitute the protuberances.
The whirlpools of the sun, like those on the earth, are
of all dimensions, from the scarcely visible pores to the
enormous spots which we see from time to time. They

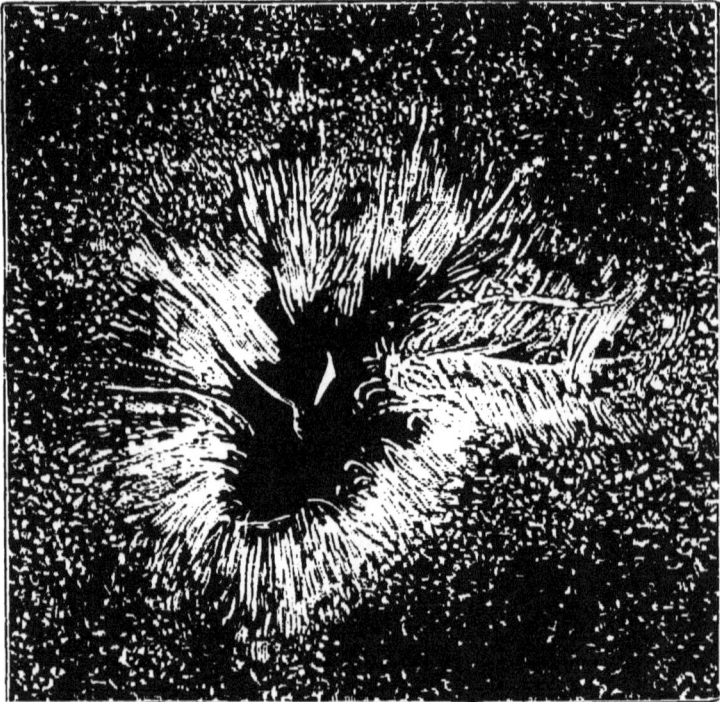

Fig. 36.—Solar Spot, by Langley.

have, like those of the earth, a marked tendency, first to
increase and then to break up, and thus form a row of
spots extending along the same parallel."

A spot of 20,000 miles diameter is quite small; there
was one 74,816 miles across, visible to the naked eye
for a week in 1843. This particular sun-spot somewhat

helped the Millerites. On the day of the eclipse, in 1858, a spot over 107,000 miles in extent was clearly seen. In such vast tempests, if there were ships built as large as the whole earth, they would be tossed like autumn leaves in an ocean storm.

The revolution of the sun carries a spot across its face in about fourteen days. After a lapse of as much more time, they often reappear on the other side, changed but recognizable. They often break out or disappear under the eye of the observer. They divide like a piece of ice dropped on a frozen pond, the pieces sliding off in every direction, or combine like separate floes driven together into a pack. Sometimes a spot will last for more than two hundred days, recognizable through six or eight revolutions. Sometimes a spot will last only half an hour.

The velocities indicated by these movements are incredible. An up-rush and down-rush at the sides has been measured of twenty miles a second; a side-rush or whirl, of one hundred and twenty miles a second. These tempests rage from a few days to half a year, traversing regions so wide that our Indian Ocean, the realm of storms, is too small to be used for comparison; then, as they cease, the advancing sides of the spots approach each other at the rate of 20,000 miles an hour; they strike together, and the rising spray of fire leaps thousands of miles into space. It falls again into the incandescent surge, rolls over mountains as the sea over pebbles, and all this for eon after eon without sign of exhaustion or diminution. All these swift succeeding Himalayas of fire, where one hundred worlds could be buried, do not usually prevent the sun's appearing to our far-off eyes as a perfect sphere.

What the Sun does for us.

To what end does this enormous power, this central source of power, exist? That it could keep all these gigantic forces within itself could not be expected. It is in a system where every atom is made to affect every other atom, and every world to influence every other. The Author of all lives only to do good, to send rain on the just and unjust, to cause his sun to rise on the evil and the good, and to give his spirit, like a perpetually widening river, to every man to profit withal.

The sun reaches his unrelaxing hand of gravitation to every other world at every instant. The tendency of every world is to fly off in a straight line. This tendency must be momentarily curbed, and the planet held in its true curve about the sun. These giant worlds must be perfectly handled. Their speed, amounting to seventy times as fast as that of a rifle-ball, must be managed. Each and every world may be said to be lifted momentarily and swung perpetually at arm's-length by the power of the sun.

The sun warms us. It would convey but a small idea of the truth to state how many hundreds of millions of cubic miles of ice could be melted by the sun every second without quenching its heat; but if any one has any curiosity to know, it is 287,200,000 cubic miles of ice per second.

We journey through space which has a temperature of 200° below zero; but we live, as it were, in a conservatory, in the midst of perpetual winter. We are roofed over by the air that treasures the heat, floored under by strata both absorptive and retentive of heat, and between the earth and air violets grow and grains ripen. The sun has a strange chemical power. It kisses the cold earth, and it blushes with flowers and matures the fruit and grain. We are feeble creatures, and the sun gives us force. By it the light winds move one-eighth of a mile an hour, the storm fifty miles, the hurricane one hundred. The force is as the square of the velocity. It is by means of the sun that the merchant's white-sailed ships are blown safely home. So the sun carries off the miasma of the marsh, the pollution of cities, and then sends the winds to wash and cleanse themselves in the sea-spray. The water-falls of the earth turn machinery, and make Lowells and Manchesters possible, because the sun lifted all that water to the hills.

Intermingled with these currents of air are the currents of electric power, all derived from the sun. These have shown their swiftness and willingness to serve man. The sun's constant force displayed on the earth is equal to 543,000,000,000 engines of 400-horse power each, working day and night; and yet the earth receives only $\frac{1}{2381000000}$ part of the whole force of the sun.

Besides all this, the sun, with provident care, has made and given to us coal. This omnipotent worker has stored away in past ages an inexhaustible reservoir of his power which man may easily mine and direct, thus releasing himself from absorbing toil.

EXPERIMENTS.

Any one may see the spots on the sun who has a spy-glass. Darken the room and put the glass through an opening toward the sun, as shown in Fig. 37. The eye-piece should be drawn out about half an inch beyond its usual focusing for distant objects. The farther it is drawn, the nearer must we hold the screen for a perfect image.

By holding a paper near the eye-piece, the proper direction of the instrument may be discovered without injury to the eyes. By this means the sun can be studied from day to day, and its spots or the transits of Mercury and Venus shown to any number of spectators.

Fig. 37.—Holding Telescope to see the Sun's Spots.

First covering the eyes with very dark or smoked glasses, erect a disk of pasteboard four inches in diameter between you and the sun; close one eye; stand near it, and the whole sun is obscured. Withdraw from it till the sun's rays just shoot over the edge of the disk on every side. Measure the distance from the eye to the disk. You will be able to determine the distance of the sun by the rule of three: thus, as four inches is to 860,000 miles, so is distance from eye to disk to distance from disk to the sun. Take such measurements at sunrise, noon, and sunset, and see the apparently differing sizes due to refraction.

VI.

THE PLANETS, AS SEEN FROM SPACE

"He hangeth the earth upon nothing."—*Job* xxvi. 7.

5

"Let a power be delegated to a finite spirit equal to the projection of the most ponderous planet in its orbit, and, from an exhaustless magazine, let this spirit select his grand central orb. Let him with puissant arm locate it in space, and, obedient to his mandate, there let it remain forever fixed. He proceeds to select his planetary globes, which he is now required to marshal in their appropriate order of distance from the sun. Heed well this distribution; for should a single globe be misplaced, the divine harmony is destroyed forever. Let us admit that finite intelligence may at length determine the order of combination; the mighty host is arrayed in order. These worlds, like fiery coursers, stand waiting the command to fly. But, mighty spirit, heed well the grand step, ponder well the direction in which thou wilt launch each waiting world; weigh well the mighty impulse soon to be given, for out of the myriads of directions, and the myriads of impulsive forces, there comes but a single combination that will secure the perpetuity of your complex scheme. In vain does the bewildered finite spirit attempt to fathom this mighty depth. In vain does it seek to resolve the stupendous problem. It turns away, and while endued with omnipotent power, exclaims, 'Give to me infinite wisdom, or relieve me from the impossible task!'"—O. M. MITCHEL, LL.D.

VI.

THE PLANETS, AS SEEN FROM SPACE.

IF we were to go out into space a few millions of miles from either pole of the sun, and were endowed with wonderful keenness of vision, we should perceive certain facts, viz: That space is frightfully dark except when we look directly at some luminous body. There is no air to bend the light out of its course, no clouds or other objects to reflect it in a thousand directions. Every star is a brilliant point, even in perpetual sunshine. The cold is frightful beyond the endurance of our bodies. There is no sound of voice in the absence of air, and conversation by means of vocal organs being impossible, it must be carried on by means of mind communication. We see below an unrevolving point on the sun that marks its pole. Ranged round in order are the various planets, each with its axis pointing in very nearly the same direction. All planets, except possibly Venus, and all moons except those of Uranus and Neptune, present their equators to the sun. The direction of orbital and axial revolution seen from above the North Pole would be opposite to that of the hands of a watch.

The speed of this orbital revolution must be proportioned to the distance from the sun. The attraction of the sun varies inversely as the square of the distance.

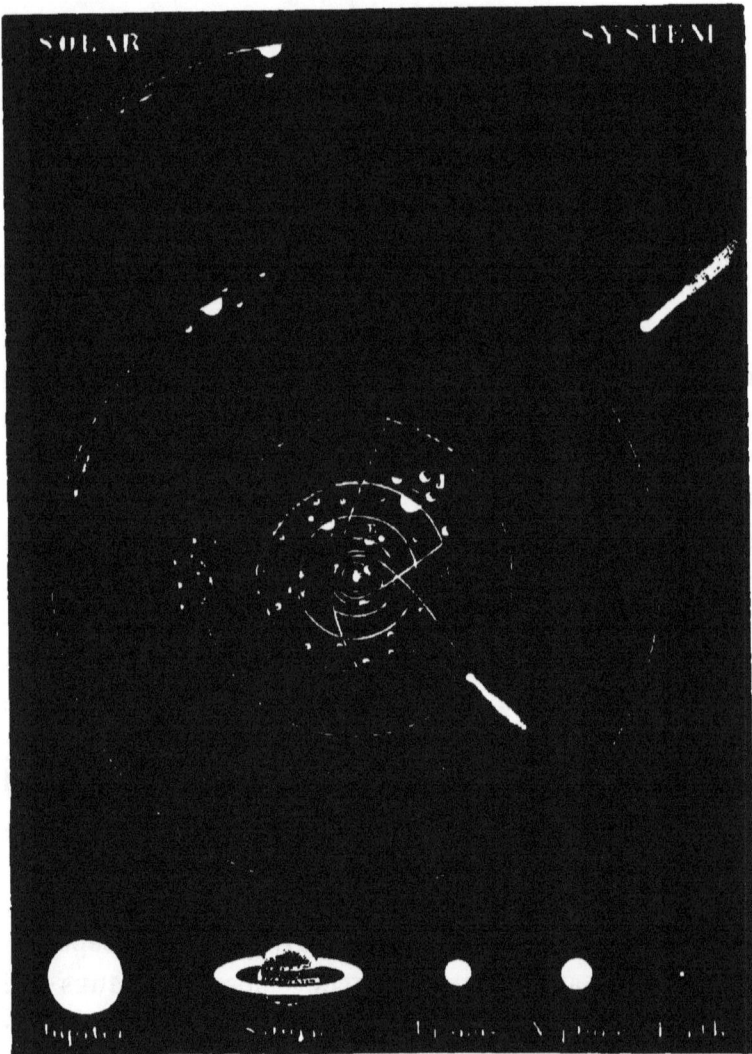

Fig. 38.—Orbits and Comparative Sizes of the Planets.

It holds a planet with a certain power; one twice as far off, with one-fourth that power. This attraction must be counterbalanced by centrifugal force; great force from great speed when attraction is great, and small from less

speed when attractive power is diminished by distance,
Hence Mercury must go 29.5 miles per second—seventy times as fast as a rifle-ball that goes two-fifths of a
mile in a second—or be drawn into the sun; while
Neptune, seventy-five times as far off, and hence attracted only $\frac{1}{5675}$ as much, must be slowed down to 3.4
miles a second to prevent its flying away from the fee-
bler attraction of the sun. The orbital velocity of the
various planets in miles per second is as follows:

Mercury	29.55	Jupiter	8.06
Venus	21.61	Saturn	5.95
Earth	18.38	Uranus	4.20
Mars	14.99	Neptune	3.36

Hence, while the earth makes one revolution in its
year, Mercury has made over four revolutions, or passed through four years; the slower Neptune has made
only $\frac{1}{164}$ of one revolution.

The time of axial rotation which determines the
length of the day varies with different planets. The
periods of the four planets nearest the sun vary only
half an hour from that of the earth, while the enormous bodies of Jupiter and Saturn revolve in ten and
ten and a quarter hours respectively. This high rate of
speed, and its resultant, centrifugal force, has aided in
preventing these bodies from becoming as dense as they
would otherwise be—Jupiter being only 0.24 as dense as
the earth, and Saturn only 0.13. This extremely rapid
revolution produces a great flattening at the poles. If
Jupiter should rotate four times more rapidly than it
does, it could not be held together compactly. As it is,
the polar diameter is five thousand miles less than the
equatorial: the difference in diameters produced by the

same cause on the earth, owing to the slower motion
and smaller mass, being only twenty-six miles. The
effect of this will be more specifically treated hereafter.

The difference in the size of the planets is very no-
ticeable. If we represent the sun by a gilded globe
two feet in diameter, we must represent Vulcan and
Mercury by mustard-seeds; Venus, by a pea; Earth, by
another; Mars, by one-half the size; Asteroids, by the
motes in a sunbeam; Jupiter, by a small-sized orange;
Saturn, by a smaller one; Uranus, by a cherry; and
Neptune, by one a little larger.

Apply the principle that attraction is in proportion
to the mass, and a man who weighs one hundred and
fifty pounds on the earth weighs three hundred and
ninety-six on Jupiter, and only fifty-eight on Mars;
while on the Asteroids he could play with bowlders for
marbles, hurl hills like Milton's angels, leap into the
fifth-story windows with ease, tumble over precipices
without harm, and go around the little worlds in seven
jumps.

The seasons of a planet are caused by the inclination
of its axis to the plane of its orbit. In Fig. 39 the rotating earth is seen at A, with its northern pole turning
in constant sunlight, and its southern pole in constant
darkness; everywhere south of the equator is more darkness than day, and hence winter. Passing on to B, the
world is seen illuminated equally on each side of the
equator. Every place has its twelve hours' darkness
and light at each revolution. But at C—the axis of the
earth always preserving the same direction—the northern pole is shrouded in continual gloom. Every place

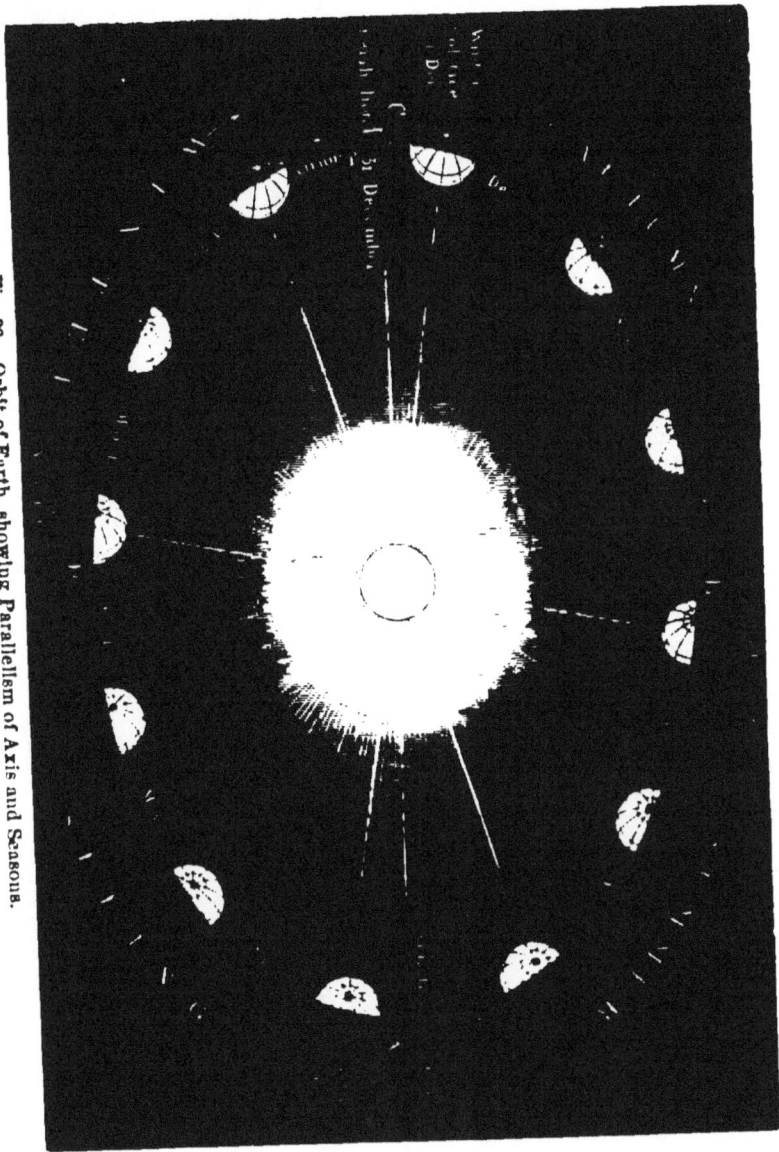

Fig. 39.—Orbit of Earth, showing Parallelism of Axis and Seasons.

north of the equator gets more darkness than light, and hence winter.

The varying inclination of the axes of the different planets gives a wonderful variety to their seasons. The sun is always nearly over the equator of Jupiter, and every place has nearly its five hours day and five hours night. The seasons of Earth, Mars, and Saturn are so much alike, except in length, that no comment is nec essary. The ice-fields at either pole of Mars are observed to enlarge and contract, according as it is winter or summer there. Saturn's seasons are each seven and a half years long. The alternate darkness and light at the poles is fifteen years long.

But the seasons of Venus present the greatest anomaly, if its assigned inclination of axis (75°) can be relied on as correct, which is doubtful. Its tropic zone extends nearly to the pole, and at the same time the winter at the other pole reaches the equator. The short period of this planet causes it to present the south pole to the sun only one hundred and twelve days after it has been scorching the one at the north. This gives two winters, springs, summers, and autumns to the equator in two hundred and twenty-five days.

If each whirling world should leave behind it a trail of light to mark its orbit, and our perceptions of form were sufficiently acute, we should see that these curves of light are not exact circles, but a little flattened into ellipses, with the sun always in one of the foci. Hence each planet is nearer to the sun at one part of its orbit than another; that point is called the perihelion, and the farthest point aphelion. This eccentricity of orbit, or distance of the sun from the centre, is very small.

In the case of Venus it is only .007 of the whole, and in no instance is it more than .2, viz., that of Mercury. This makes the sun appear twice as large, bright, and hot as seen and felt on Mercury at its perihelion than at its aphelion. The earth is 3,236,000 miles nearer to the sun in our winter than summer. Hence the summer in the southern hemisphere is more intolerable than in the northern. But this eccentricity is steadily diminishing at a uniform rate, by reason of the perturbing influence of the other planets. In the case of some other planets it is steadily increasing, and, if it were to go on a sufficient time, might cause frightful extremes of temperature; but Lalande has shown that there are limits at which it is said, "Thus far shalt thou go, and no farther." Then a compensative diminution will follow.

Conceive a large globe, to represent the sun, floating in a round pond. The axis will be inclined 7½° to the surface of the water, one side of the equator be 7½° below the surface, and the other side the same distance above. Let the half-submerged earth sail around the sun in an appropriate orbit. The surface of the water will be the plane of the orbit, and the water that reaches out to the shore, where the stars would be set, will be the plane of the ecliptic. It is the plane of the earth's orbit extended to the stars.

The orbits of all the planets do not lie in the same plane, but are differently inclined to the plane of the ecliptic, or the plane of the earth's orbit. Going out from the sun's equator, so as to see all the orbits of the planets on the edge, we should see them inclined to that of the earth, as in Fig. 40.

If the earth, and Saturn, and Pallas were lying at

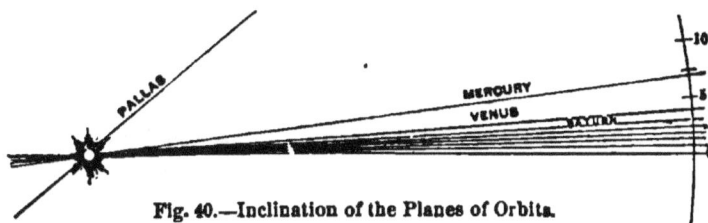

Fig. 40.—Inclination of the Planes of Orbits.

right angles with the nodal line of their orbits, and in the same direction from the sun, and the outer bodies were to start in a direct line for the sun, they would not collide with the earth on their way; but Saturn would pass 4,000,000 and Pallas 50,000,000 miles over our heads. From this same cause we do not see Venus and Mercury make a transit across the disk of the sun at every revolution.

Fig. 41 shows a view of the orbits of the earth and

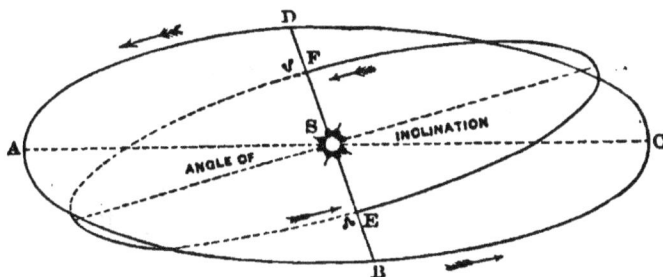

Fig. 41.—Inclination of Orbits of Venus and Earth. Nodal Line, D B.

Venus seen not from the edge but from a position somewhat above. The point E, where Venus crosses the plane of the earth's orbit, is called the ascending node. If the earth were at B when Venus is at E, Venus would be seen on the disk of the sun, making a transit. The same would be true if the earth were at D, and Venus at the descending node F.

This general view of the flying spheres is full of interest. While quivering themselves with thunderous noises, all is silent about them; earthquakes may be struggling on their surfaces, but there is no hint of contention in the quiet of space. They are too distant from one another to exchange signals, except, perhaps, the fleet of asteroids that sail the azure between Mars and Jupiter. Some of these come near together, continuing to fill each other's sky for days with brightness, then one gradually draws ahead. They have all phases for each other—crescent, half, full, and gibbous. These hundreds of bodies fill the realm where they are with inexhaustible variety. Beyond are vast spaces—cold, dark, void of matter, but full of power. Occasionally a little spark of light looms up rapidly into a world so huge that a thousand of our earths could not occupy its vast bulk. It swings its four or eight moons with perfect skill and infinite strength; but they go by and leave the silence unbroken, the darkness unlighted for years. Nevertheless, every part of space is full of power. Nowhere in its wide orbit can a world find a place; at no time in its eons of flight can it find an instant when the sun does not hold it in safety and life.

The Outlook from the Earth.

If we come in from our wanderings in space and take an outlook from the earth, we shall observe certain movements, easily interpreted now that we know the system, but nearly inexplicable to men who naturally supposed that the earth was the largest, most stable, and central body in the universe.

We see, first of all, sun, moon, and stars rise in the east, mount the heavens, and set in the west. As I revolve in my pivoted study-chair, and see all sides of the room — library, maps, photographs, telescope, and windows—I have no suspicion that it is the room that whirls; but looking out of a car-window in a depot at another car, one cannot tell which is moving, whether it be his car or the other. In regard to the world, we have come to feel its whirl. We have noticed the pyramids of Egypt lifted to hide the sun; the mountains of Hymettus hurled down, so as to disclose the moon that was behind them to the watchers on the Acropolis; and the mighty mountains of Moab removed to reveal the stars of the east. Train the telescope on any star; it must be moved frequently, or the world will roll the instrument away from the object. Suspend a cannon-ball by a fine wire at the equator; set it vibrating north and south, and it swings all day in precisely the same direction. But suspend it directly over the north pole, and set it swinging toward Washington; in five hours after it is swinging toward the Sandwich Islands; in twelve hours, toward Siam, in Asia; in eighteen hours, toward Rome, in Italy; and in twenty-four, toward Washington again, not because it has changed the plane of its vibration, but because the earth has whirled beneath it, and the torsion of the wire has not been sufficient to compel the plane of the original direction to change with the turning of the earth. The law of inertia keeps it moving in the same direction. The same experimental proof of revolution is shown in a proportional degree at any point between the pole and the equator.

But the watchers on the Acropolis do not get turned over so as to see the moon at the same time every night.

We turn down our eastern horizon, but we do not find
fair Luna at the same moment we did the night before.
We are obliged to roll on for some thirty to fifty minutes longer before we find the moon. It must be going in the same direction, and it takes us longer to get
round to it than if it were always in the same spot; so
we notice a star near the moon one night—it is 13° west
of the moon the next night. The moon is going around

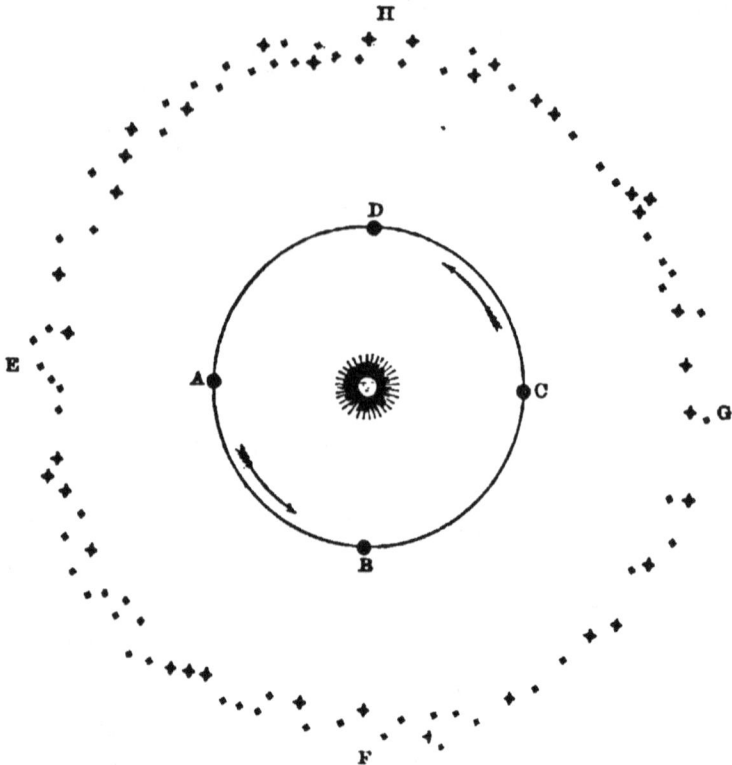

Fig. 42.—Showing the Sun's Movement among the Stars.

the earth from west to east, and if it goes 13° in one
day, it will take a little more than twenty-seven days to
go the entire circle of 360°.

In our outlook we soon observe that we do not by our revolution come to see the same stars rise at the same hour every night. Orion and the Pleiades, our familiar friends in the winter heavens, are gone from the summer sky. Have they fled, or are we turned from them? This is easily understood from Fig. 42.

When the observer on the earth at A looks into the midnight sky he sees the stars at E; but as the earth passes on to B, he sees those stars at E four minutes sooner every night; and at midnight the stars at F are over his head. Thus in a year, by going around the sun, we have every star of the celestial dome in our midnight sky. We see also how the sun appears among the successive constellations. When we are at A, we see the sun among the stars at G; but as we move toward B, the sun appears to move toward H. If we had observed the sun rise on the 20th of August, 1876, we should have seen it rise a little before Regulus, and a little south of it, in such a relation as circle 1 is to the star in Fig. 43. By sunset the earth had moved enough to make the sun appear to be at circle 2, and by the next morning at circle 3, at which time Regulus would rise before the sun. Thus the earth's motion seems to make the sun traverse a regular circle among the stars once a year: but it is not the sun that moves.

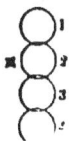

Fig. 43.

There are certain stars that have such irregular, uncertain, vagarious ways that they were called vagabonds, or planets, by the early astronomers. Here is the path of Jupiter in the year 1866 (Fig. 44). These bodies go forward for awhile, then stop, start aside, then retrograde, and go on again. Some are never seen far from the sun, and others in all parts of the ecliptic.

Fig. 44.

First see them as they stand to-day, as in Fig. 45. The observer stands on the earth at A. It has rolled over so far that he cannot see the sun; it has set. But

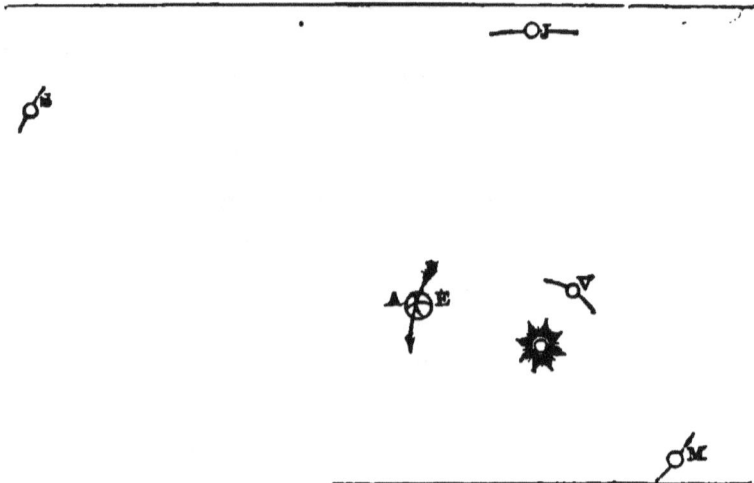

Fig. 45.—Showing Position of Planets.

Venus is still in sight; Jupiter is 45° behind Venus, and Saturn is seen 90° farther east. When A has roll-ed a little farther, if he is awake, he will see Mars be-fore he sees the sun; or, in common language, Venus will set after, and Mars rise before the sun. All these bodies at near and far distances seem set in the starry dome, as the different stars seem in Fig. 42, p. 110.

The mysterious movements of advance and retreat are rendered intelligible by Fig. 46. The planet Mer-cury is at A, and, seen from the earth, B is located at *a*,

on the background of the stars it seems to be among.
It remains apparently stationary at *a* for some time, be-

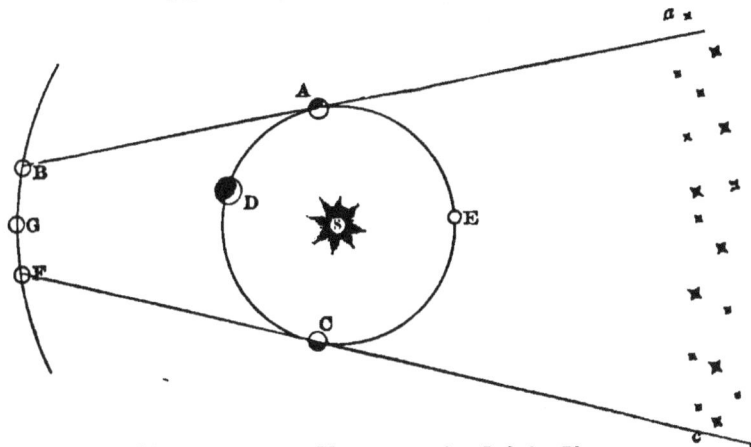

Fig. 46.—Apparent Movements of an Inferior Planet.

cause approaching the earth in nearly a straight line.
Passing D to C, it appears to retrograde among the
stars to *c;* remains apparently stationary for some time,
then, in passing from C to E and A, appears to pass
back among the stars to *a*. The progress of the earth,
meanwhile, although it greatly retards the apparent motion from A to C, greatly hastens it from C to A.

It is also apparent that Mercury and Venus, seen
from the earth, can never appear far from the sun.
They must be just behind the sun as evening stars, or
just before it as heralds of the morning. Venus is never more than 47° from the sun, and Mercury never more
than 30°; indeed, it keeps so near the sun that very
few people have ever seen the brilliant sparkler. Observe how much larger the planet appears near the earth
in conjunction at D than in opposition at E. Observe
also what phases it must present, and how transits sometimes take place.

The movement of a superior planet, one whose orbit is exterior to the earth, is clear from Fig. 47. When the earth is at A and Mars at B, it will appear among the stars at C. When the earth is at D, Mars having moved more slowly to E, will have retrograded to F. It remains there while the earth passes on, in a line near-ly straight, from Mars to G; then, as the earth begins to curve around the sun, Mars will appear to retraverse

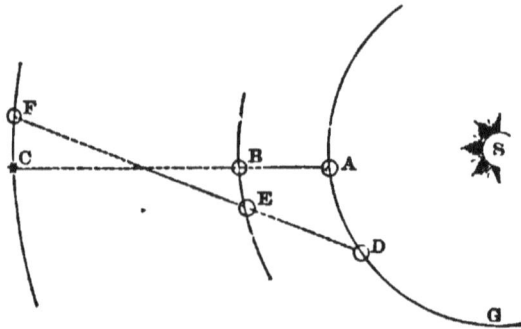

Fig. 47.—Illustrating Movements of a Superior Planet.

the distance from F to C, and beyond. The farther the superior planet is from the earth the less will be the retrograde movement.

The reader should draw the orbits in proportion, and, remembering the relative speed of each planet, note the movement of each in different parts of their orbits.

To account for these most simple movements, the earlier astronomers invented the most complex and im-possible machinery. They thought the earth the centre, and that the sun, moon, and stars were carried about it, as stoves around a person to warm him. They thought these strange movements of the planets were accom-plished by mounting them on subsidiary eccentric wheels in the revolving crystal sphere. All that was needed to give them a right conception was a sinking of their world and themselves to an appropriate proportion, and an enlargement of their vision, to take in from an exalted stand-point a view of the simplicity of the perfect plan.

EXPERIMENTS.

Fix a rod, or tube, or telescope pointing at a star in the east or west, and the earth's revolution will be apparent in a moment, turning the tube away from the star. Point it at stars about the north pole, and those on one side will be found going in an opposite direction from those on the other, and very much slower than those about the equator. Any one can try the pendulum experiment who has access to some lofty place from which to suspend the ball. It was tried in Bunker Hill Monument a few years ago, and is to be tried in Paris, in the summer of 1879, with a seven-hundred-pound pendulum and a suspending wire seventy yards long. The advance and retrograde movements of planets can be illustrated by two persons walking around a centre and noticing the place where the person appears projected on the wall beyond.

PROCESSION OF STARS AND SOULS.

I STOOD upon the open casement,
  And looked upon the night,
And saw the westward-going stars
  Pass slowly out of sight.

Slowly the bright procession
  Went down the gleaming arch,
And my soul discerned the music
  Of the long triumphal march;

Till the great celestial army,
  Stretching far beyond the poles,
Became the eternal symbol
  Of the mighty march of souls.

Onward, forever onward,
  Red Mars led on his clan ;
And the moon, like a mailèd maiden,
  Was riding in the van.

And some were bright in beauty,
  And some were faint and small,
But these might be, in their great heights,
  The noblest of them all.

Downward, forever downward,
  Behind earth's dusky shore,
They passed into the unknown night—
  They passed, and were no more.

No more! oh, say not so!
  And downward is not just ;
For the sight is weak and the sense is dim
  That looks through heated dust.

The stars and the mailèd moon,
  Though they seem to fall and die,
Still sweep in their embattled lines
  An endless reach of sky.

And though the hills of Death
  May hide the bright array,
The marshalled brotherhood of souls
  Still keeps its onward way.

Upward, forever upward,
  I see their march sublime,
And hear the glorious music
  Of the conquerors of Time.

And long let me remember
  That the palest fainting one
May to diviner vision be
  A bright and blazing sun.

              Thomas Buchanan Read.

VII.

SHOOTING-STARS, METEORS, AND COMETS.

"The Lord cast down great stones from heaven upon them unto Aze kah, and they died."—*Joshua* x. 11.

Their orbits are all parallel. Those coming in direct line to the eye appear as stars, having no motion. Those at one side of this line are seen in foreshortened perspective. Those farthest from the centre, other things being equal, appear longest. The centre, called the radiant point, of these November meteors is situated in Leo; that of the August meteors in Perseus. Over fifty such radiant points have been discovered. Over 30,000 meteors have been visible in an hour.

VII.

SHOOTING-STARS, METEORS, AND COMETS.

BEFORE particularly considering the larger aggrega-
tions of matter called planets or worlds as individuals,
it is best to investigate a part of the solar system consisting of smaller collections of matter scattered everywhere through space. They are of various densities,
from a cloudlet of rarest gas to solid rock; of various
sizes, from a grain's weight to little worlds; of various relations to each other, from independent individuality to related streams millions of miles long. When
they become visible they are called shooting-stars, which
are evanescent star-points darting through the upper
air, leaving for an instant a brilliant train; meteors,
sudden lights, having a discernible diameter, passing
over a large extent of country, often exploding with
violence (Fig. 48), and throwing down upon the earth
aerolites; and comets, vast extents of ghostly light,
that come we know not whence and go we know not
whither. All these forms of matter are governed by
the same laws as the worlds, and are an integral part
of the whole system—a part of the unity of the universe.

Every one has seen the so-called shooting-stars.
They break out with a sudden brilliancy, shoot a few
degrees with quiet speed, and are gone before we can
say, "See there!" The cause of their appearance, the

conversion of force into heat by their contact with our atmosphere, has been already explained. Other facts remain to be studied. They are found to appear about seventy-three miles above the earth, and to disappear

Fig. 48.—Explosion of a Bolide.

about twenty miles nearer the surface. Their average velocity, thirty-five, sometimes rises to one hundred miles a second. They exhibit different colors, according to their different chemical substances, which are consumed. The number of them to be seen on different nights is exceedingly variable; sometimes not more than five or six an hour, and sometimes so many that a man cannot count those appearing in a small section of sky. This variability is found to be periodic. There are everywhere in space little meteoric masses of matter, from the weight of a grain to a ton, and from the density of gas to rock. The earth meets 7,500,000 little bodies every day—there is collision—the little meteoroid gives out its lightning sign of extinction, and, consumed in fervent heat, drops to the earth as gas or dust. If we add the number light enough to be seen by a telescope, they cannot be less than 400,000,000 a day. Everywhere we go, in a space as large as that occupied by the earth and its atmosphere, there must be at least 13,000 bodies—one in 20,000,000 cubic miles —large enough to make a light visible to the naked eye, and forty times that number capable of revealing themselves to telescopic vision. Professor Peirce is

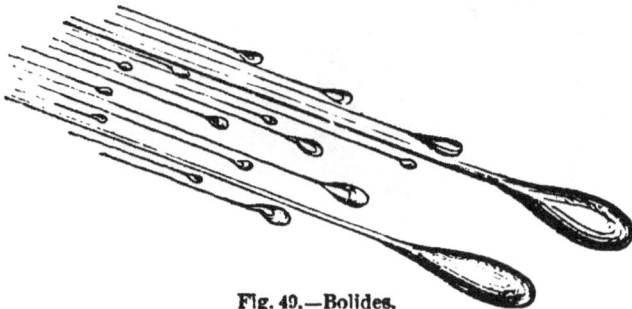

Fig. 49.—Bolides.

about to publish, as the startling result of his investigations, "that the heat which the earth receives directly from meteors is the same in amount which it receives from the sun by radiation, and that the sun receives five-sixths of its heat from the meteors that fall upon it." But this is not received by other astronomers.

6

In 1783 Dr. Schmidt was fortunate enough to have a telescopic view of a system of bodies which had turned into meteors. These were two larger bodies followed by several smaller ones, going in parallel lines till they were extinguished. They probably had been revolving about each other as worlds and satellites be-

Fig. 50.—Santa Rosa Aerolite.

fore entering our atmosphere. It is more than probable that the earth has many such bodies, too small to be visible, revolving around it as moons.

Aerolites.

Sometimes the bodies are large enough to bear the heat, and the unconsumed centre comes to the earth.

Their velocity has been lessened by the resisting air, and the excessive heat diminished. Still, if found soon after their descent, they are too hot to be handled. These are called aerolites or air-stones. There was a fall in Iowa, in February, 1875, from which fragments amounting to five hundred pounds weight were secured. On the evening of December 21st, 1876, a meteor of unusual size and brilliancy passed over the states of Kansas, Missouri, Illinois, Indiana, and Ohio. It was first seen in the western part of Kansas, at an altitude of about sixty miles. In crossing the State of Missouri it began to explode, and this breaking up continued while passing Illinois, Indiana, and Ohio, till it consisted of a large flock of brilliant balls chasing each other across the sky, the number being variously estimated at from twenty to one hundred. It was accompanied by terrific explosions, and was seen along a path of not less than a thousand miles. When first seen in Kansas, it is said to have appeared as large as the full moon, and with a train from twenty-five to one hundred feet long. Another, very similar in appearance and behavior, passed over a part of the same course in February, 1879. At Laigle, France, on April 26th, 1803, about one o'clock in the day, from two to three thousand fell. The largest did not exceed seventeen pounds weight. One fell in Weston, Connecticut, in 1807, weighing two hundred pounds. A very destructive shower is mentioned in the book of Joshua, chap. x. ver. 11.

These bodies are not evenly distributed through space. In some places they are gathered into systems which circle round the sun in orbits as certain as those of the planets. The chain of asteroids is an illustration of meteoric bodies on a large scale. They are hundreds in number — meteors are millions. They have their region of travel, and the sun holds them and the giant Jupiter by the same power. The Power that cares for a world cares for a sparrow. If their orbit so lies that a planet passes through it, and the planet and the meteors are at the point of intersection at the same time, there must be collisions, and the lightning signs of extinction proportioned to the number of little bodies in a given space.

It is demonstrated that the earth encounters more than one hundred such systems of meteoric bodies in a single year. It passes through one on the 10th of August, another on the 11th of November. In a certain part of the first there is an agglomeration of bodies sufficient to become visible as it approaches the sun, and this is known as the comet of 1862; in the second is a similar agglomeration, known as Temple's comet. It is repeating the same thing to say that meteoroids follow in the train of the comets. The probable orbit of the November meteors and the comet of 1866 is an exceedingly elongated ellipse, embracing the orbit of the earth at one end and a portion of the orbit of Uranus at the other (Fig. 51). That of the August meteors and the comet of 1862 embraces the orbit of the earth at one end, and thirty per cent. of the other end is beyond the orbit of Neptune.

In January, 1846, Biela's comet was observed to be divided. At its next return, in 1852, the parts were 1,500,000 miles apart. They could not be found on their periodic returns in 1859, 1865, and 1872; but it should have crossed the earth's orbit early in September, 1872. The earth itself would arrive at the point of crossing two or three months later. If the law of revolution held, we might still expect to find some of the trailing meteoroids of the comet not gone by on our ar-

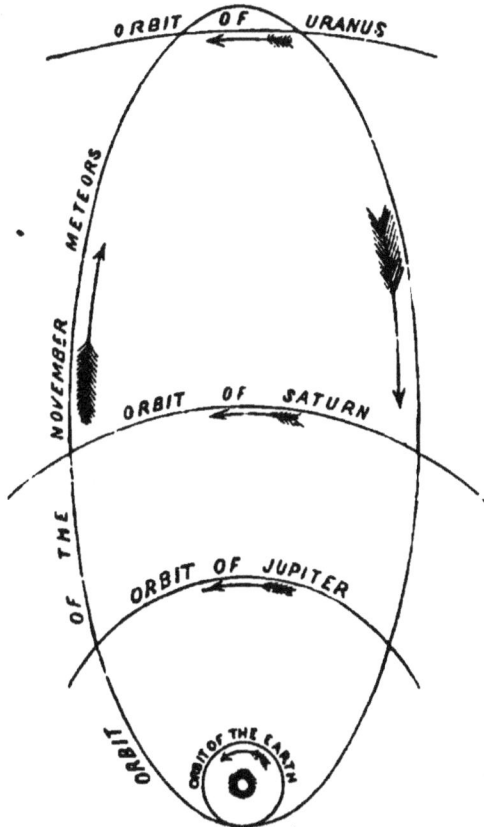

Fig. 51.—Orbit of the November Meteors and the Comet of 1866.

rival. It was shown that the point of the earth that would strike them would be toward a certain place in the constellation of Andromeda, if the remains of the diluted comet were still there. The prediction was verified in every respect. At the appointed time, place, and direction, the streaming lights were in our sky. That these little bodies belonged to the original comet none can doubt. By the perturbations of planetary attraction, or by different original velocities, a comet may be lengthened into an invisible stream, or an invisible stream agglomerated till it is visible as a comet.

Comets.

Comets will be most easily understood by the foregoing considerations. They are often treated as if they were no part of the solar system; but they are under the control of the same laws, and owe their existence, motion, and continuance to the same causes as Jupiter and the rest of the planets. They are really planets of wider wandering, greater ellipticity, and less density. They have periodic times less than asteroids, and fifty times as great as Neptune. They are little clouds of gas or meteoric matter, or both, darting into the solar system from every side, at every angle with the plane of the ecliptic, becoming luminous with reflected light, passing the sun, and returning again to outer darkness. Sometimes they have no tail, having a nucleus surrounded by nebulosity like a dim sun with zodiacal light; sometimes one tail, sometimes half a dozen. These follow the comet to perihelion, and precede it afterward (Fig. 52). The orbits of some comets are enormously elongated; one end may lie inside the earth's orbit, and the other end be as far beyond Neptune as that is from the sun. Of course only a small part of such a curve can be studied by us: the comet is visible only when near the sun. The same curve around the sun may be an orbit that will bring it back again,

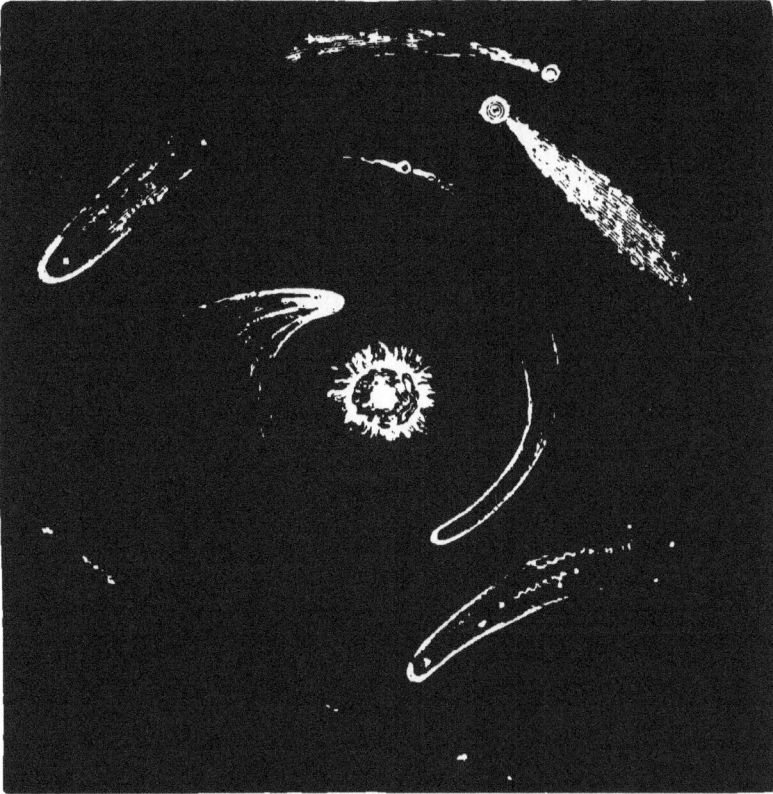

Fig. 52.—Aspects of Remarkable Comets.

or one that will carry it off into infinite space, never to
return. One rate of speed on that curve indicates an
elliptical orbit; it returns; a greater rate of speed indicates that it will take a parabolic orbit, and never
return. The exact rate of speed is exceedingly difficult
to determine; hence it cannot be confidently asserted
that any comet ever visible will not return. They may
all belong to the solar system; but some will certainly
be gone thousands of years before their fiery forms will
greet the watchful eyes of dwellers on the earth. A
comet that has an elliptic orbit may have it changed to

parabolic by the accelerations of its speed, by attracting planets; or a parabolic comet may become elliptic, and so permanently attracted to the system by the retardations of attracting bodies. A comet of long period may be changed to one of short period by such attraction, or *vice versa.* Orbits may be changed without affecting speed.

The number of comets, like that of meteor streams, is exceedingly large. Five hundred have been visible to the naked eye since the Christian era. Two hundred have been seen by telescopes invented since their invention. Some authorities estimate the number belonging to our solar system by millions; Professor Peirce says more than five thousand millions.

Famous Comets.

The comet of 1680 is perhaps the one that appeared in A.D. 44, soon after the death of Julius Cæsar, also in the reign of Justinian, A.D. 531, and in 1106. This is not determined by any recognizable resemblance. It had a tail 70° long; it was not all arisen when its head reached the meridian. It is possible, from the shape of its orbit, that it has a periodic time of nine thousand years, or that it may have a parabolic orbit, and never return. Observations taken two hundred years ago have not the exactness necessary to determine so delicate a point.

On August 19th, 1682, Halley discovered a comet which he soon declared to be one seen by Kepler in 1607. Looking back still farther, he found that a comet was seen in 1531 having the same orbit. Still farther, by the same exact period of seventy-five years, he found that it was the same comet that had disturbed the equanimity of Pope Calixtus in 1456. Calculations were undertaken as to the result of all the accelerations and retardations by the attractions of all the planets for the next seventy-five years. There was not time to finish all the work; but a retardation of six hundred and eighteen days was determined, with a possible error of thirty days. The comet actually came to time within thirty-three days, on March 12th, 1759. Again its return was calculated with more laborious care. It came to time and passed the sun within three days of the predicted time, on the 16th of November, 1835. It passed from sight of the most powerful telescopes the following May, and has never since been seen by human eye. But the eye of science sees it as having passed its aphelion beyond the orbit of Neptune in 1873, and is already hastening back to the warmth and light of the sun. It will be looked for in 1911; and there is good hope of predicting, long before it is seen, the time of its perihelion within a day.

Biela's lost Comet.—This was a comet with a periodic time of six years and eight months. It was observed in January, 1846, to have separated into two parts of unequal brightness. The lesser part grew for a month until it equalled the other, then became smaller and disappeared, while the other was visible a month longer. At disappearance the parts were 200,000 miles asunder. On its next return, in 1852, the parts were 1,500,000 miles apart; sometimes one was brighter and sometimes the other; which was the fragment and which was the main body could not be recognized. They vanished in September, 1852, and have never been seen since. Three revolutions have been made since that time, but no

6*

trace of it could be discovered. Probably the same influence that separated it into parts, separated the particles till too thin and tenuous to be seen. There is
ground for believing that the earth passed through a
part of it, as before stated under the head of meteors.

The Great Comet of 1843 passed nearer the sun than
any known body. It almost grazed the sun. If it ever
returns, it will be in A.D. 2373.

Donati's Comet of 1858.—This was one of the most
magnificent of modern times. During the first three
months it showed no tail, but from August to October
it had developed one forty degrees in length. Its period
is about two thousand years. Every reader remembers
the comet of the summer of 1874.

Encke's Comet.—This comet has become famous for
its supposed confirmation of the theory that space was
filled with a substance infinitely tenuous, which resisted
the passage of this gaseous body in an appreciable degree, and in long ages would so retard the motion of all
the planets that gravitation would draw them all one
by one into the sun. We must not be misled by the
term retardation to suppose it means behind time, for a
retarded body is before time. If its velocity is diminished, the attraction of the sun causes it to take a smaller orbit, and smaller orbits mean increased speed—hence
the supposed retardation would shorten its periodic
time. This comet was thought to be retarded two and
a half hours at each revolution. If it was, it would not
prove the existence of the resisting medium. Other
causes, unknown to us, might account for it. Subsequent and more exact calculations fail to find any retardations in at least two revolutions between 1865 and

1871. Indications point to a retardation of one and a half hours both before and since. But such discrepancy of result proves nothing concerning a resisting medium, but rather is an argument against its existence. Besides, Faye's comet, in four revolutions of seven years each, shows no sign of retardation.

The truth may be this, that a kind of atmosphere exists around the sun, perhaps revealed by the zodiacal light, that reaches beyond where Encke's comet dips inside the orbit of Mercury, and thus retards this body, but does not reach beyond the orbit of Mars, where Faye's comet wheels and withdraws.

Of what do Comets consist?

The unsolved problems pertaining to comets are very numerous and exceedingly delicate. Whence come they? Why did they not contract to centres of nebulæ? Are there regions where attractions are balanced, and matter is left to contract on itself, till the movements of suns and planets adds or diminishes attractive force on one side, and so allows them to be drawn slowly toward one planet, and its sun, or another? There is ground for thinking that the comet of 1866 and its train of meteors, visible to us in November, was thus drawn into our system by the planet Uranus. Indeed, Leverrier has conjecturally fixed upon the date of A.D. 128 as the time when it occurred; but another and closer observation of its next return, in 1899, will be needed to give confirmation to the opinion. Our sun's authority extends at least half-way to the nearest fixed star, one hundred thousand times farther than the orbit of the earth. Meteoric and cometary matter lying there, in a spherical shell about the solar system, balanced between the attraction of different suns, finally feels the power that determines its destiny toward our sun. It would take 167,000,000 years to come thence to our system.

The conditions of matter with which we are acquainted do not cover all the ground presented by these mysterious visitors. We know a gas sixteen times as light as air, but hydrogen is vastly too heavy and dense; for we see the faintest star through thousands of miles of cometary matter; we know that water may become cloudy vapor, but a little of it obscures the vision. Into what more ethereal, and we might almost say spiritual, forms matter may be changed we cannot tell. But if we conceive comets to be only gas, it would expand indefinitely in the realms of space, where there is no force of compression but its own. We might say that comets are composed of small separate masses of matter, hundreds of miles apart; and, looking through thousands of miles of them, we see light enough reflected from them all to seem continuous. Doubtless that is sometimes the case. But the spectroscope shows another state of things: it reveals in some of these comets an incandescent gas—usually some of the combinations of carbon. The conclusion, then, naturally is that there are both gas and small masses of matter, each with an orbit of its own nearly parallel to those of all the others, and that they afford some attraction to hold the mass of intermingled and confluent gas together. Our best judgment, then, is that the nucleus is composed of separate bodies, or matter in a liquid condition, capable of being vaporized by the heat of the sun, and driven off, as steam from a locomotive, into a tail. Indications of this are found in the fact that tails grow smaller at successive returns, as the matter capable of such vaporization becomes condensed. In some instances, as in that of the comet of 1843, the head was diminished by the manufacture of a tail. On the other hand, Professor Peirce showed that the nucleus of the comets of 1680, 1843, and 1858 must have had a tenacity equal to steel, to prevent being pulled apart by the tidal forces caused by their terrible perihelion sweep around the sun.

It is likely that there are great varieties of condition in different comets, and in the same comet at times. We see them but a few days out of the possible millions of their periodic time; we see them only close to the sun, under the spur of its tremendous attraction and terrible heat. This gives us ample knowledge of the path of their orbit and time of their revolution, but little ground for judgment of their condition, when they slowly round the uttermost cape of their far-voyaging, in the terrible cold and darkness, to commence their homeward flight. The unsolved problems are not all in the distant sun and more distant stars, but one of them is carried by us, sometimes near, sometimes far off; but our acquaintance with the possible forms and conditions of matter is too limited to enable us to master the difficulties.

Will Comets strike the Earth?

Very likely, since one or two have done so within a recent period. What will be the effect? That depends on circumstances. There is good reason to suppose we passed through the tail of a comet in 1861, and the only observable effect was a peculiar phosphorescent mist. If the comet were composed of small meteoric masses a brilliant shower would be the result. But if we fairly encountered a nucleus of any considerable mass and solidity, the result would be far more serious. The mass of Donati's comet has been estimated by M. Faye to be $\frac{1}{30000}$ of that of the earth. If this amount of matter were dense as water, it would make a globe five hundred miles in diameter; and if as dense as Professor Peirce proved the nucleus of this comet to be, its impact with the earth would develop heat enough to melt and vaporize the hardest rocks. Happily there is little fear of this: as Professor Newcomb says, "So small is the earth in comparison with celestial space, that if one were to shut his eyes and fire at random in the air, the chance of bringing down a bird would be better than that of a comet of any kind striking the earth." Besides, we are not living under a government of chance, but under that of an Almighty Father, who upholdeth all things by the word of his power; and no world can come to ruin till he sees that it is best.

VIII.

THE PLANETS AS INDIVIDUALS.

"Through faith we understand that the worlds [plural] were framed by the word of God, so that things which are seen were not made of things which do appear."—*Heb.* xi. 3.

"O rich and various man! thou palace of sight and sound, carrying in thy senses the morning, and the night, and the unfathomable galaxy; in thy brain the geometry of the city of God; in thy heart the power of love, and the realms of right and wrong. An individual man is a fruit which it costs all the foregoing ages to form and ripen. He is strong, not to do but to live; not in his arms, but in his heart; not as an agent, but as a fact."—EMERSON.

VIII.

THE PLANETS AS INDIVIDUALS.

How many bodies there may be revolving about the sun we have no means to determine or arithmetic to express. When the new star of the American Republic appeared, there were but six planets discovered. Since then three regions of the solar system have been explored with wonderful success. The outlying realms beyond Saturn yielded the planet Uranus in 1781, and Neptune in 1846. The middle region between Jupiter and Mars yielded the little planetoid Ceres in 1801, Pallas in 1802, and two hundred and fifty others since. The inner region between Mercury and the sun is of necessity full of small meteoric bodies; the question is, are there any bodies large enough to be seen?

The same great genius of Leverrier that gave us Neptune from the observed perturbations of Uranus, pointed out perturbations in Mercury that necessitated either a planet or a group of planetoids between Mercury and the sun. Theoretical astronomers, aided by the fact that no planet had certainly been seen, and that all asserted discoveries of one had been by inexperienced observers, inclined to the belief in a group, or that the disturbance was caused by the matter reflecting the zodiacal light.

When the total eclipse of the sun occurred in 1878,

astronomers were determined that the question of the existence of an intra-mercurial planet should be settled. Maps of all the stars in the region of the sun were carefully studied, sections of the sky about the sun were assigned to different observers, who should attend to nothing but to look for a possible planet. It was then claimed that Professor Watson and Lewis Swift, the famous comet-finder, each discovered two small bodies—four in all—within the orbit of Mercury. But Professor Peters has shown that if Professor Watson made an error of $\frac{1}{130}$ of an inch in marking, with a lead-pencil, in the darkness of the eclipse, the objects were well known fixed stars.

MERCURY.

The swift messenger of the gods; sign ☿, his caduceus.

Distance from the sun, 35,750,000 miles. Diameter, 2992 miles. Orbital revolution, 87.97 days. Orbital velocity, 1773 miles per minute. Axial revolution, unknown.

Mercury shines with a white light nearly as bright as Sirius; is always near the horizon. When nearly between us and the sun, as at D (Fig. 46, p. 113), its illuminated side nearly opposite to us, we, looking from E, see only a thin crescent of its light. When it is at its greatest angular distance from the sun, as A or C, we see it illuminated like the half-moon. When it is beyond the sun, as at E, we see its whole illuminated face like the full-moon.

The variation of its apparent size from the varying distance is very striking. At its extreme distance from the earth it subtends an angle of only five seconds; nearest to us, an angle of twelve seconds. Its distance from the earth varies nearly as one to three, and its apparent size in the inverse ratio.

When Mercury comes between the earth and the sun,
near the line where the planes of their orbits cut each
other by reason of their inclination, the dark body of
Mercury will be seen on the bright surface of the sun.
This is called a transit. If it goes across the centre of
the sun it may consume eight hours. It goes 100,000
miles an hour, and has 860,000 miles of disk to cross.
The transit of 1878 occupied seven and a half hours.
The transits for the remainder of the century will
occur:

November 7th.......... 1881	November 10th....... 1894
May 9th 1891	November 4th........ 1901

VENUS.

Goddess of beauty; its sign ♀, a mirror.

**Distance from the sun, 66,750,000 miles. Diameter, 7660
miles. Orbital Velocity, 1296 miles per minute. Axial revolution, 23h. 21m. Orbital revolution, 224.7 days.**

This brilliant planet is often visible in the daytime.
I was once delighted by seeing Venus looking down, a
little after mid-day, through the open space in the dome
of the Pantheon at Rome. It has never since seemed
to me as if the home of all the gods was deserted.
Phœbus, Diana, Venus and the rest, thronged through
that open upper door at noon of night or day. Arago
relates that Bonaparte, upon repairing to Luxemburg
when the Directory was about to give him a *fête*, was
much surprised at seeing the multitude paying more
attention to the heavens above the palace than to him
or his brilliant staff. Upon inquiry, he learned that
these curious persons were observing with astonishment
a star which they supposed to be that of the conqueror
of Italy. The emperor himself was not indifferent when

his piercing eye caught the clear lustre of Venus smiling upon him at mid-day.

This unusual brightness occurs when Venus is about five weeks before or after her inferior conjunction, and also nearest overhead by being north of the sun. This last circumstance occurs once in eight years, and came on February 16th, 1878.

Vénus may be as near the earth as 22,000,000 miles, and as far away as 160,000,000. This variation of its distances from the earth is obviously much greater than that of Mercury, and its consequent apparent size much more changeable. Its greatest and least apparent sizes are as ten and sixty-five (Fig. 53).

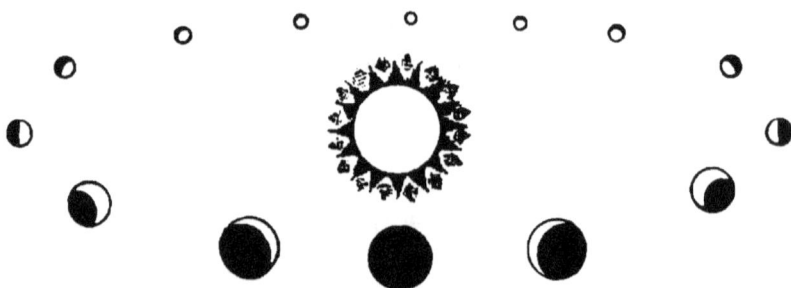

Fig. 53.—Phases of Venus, and Various Apparent Dimensions.

When Copernicus announced the true theory of the solar system, he said that if the inferior planets could be clearly seen they would show phases like the moon. When Galileo turned the little telescope he had made on Venus, he confirmed the prophecy of Copernicus. Desiring to take time for more extended observation, and still be able to assert the priority of his discovery, he published the following anagram, in which his discovery was contained:

" Hæc immatura a me jam frustra leguntur o. y."
(These unripe things are now vainly gathered by me.)

He first saw Venus as gibbous; a few months revealed it as crescent, and then he transposed his anagram into:

" Cynthiæ figuras æmulatur mater amorum."
(The mother of loves imitates the phases of Cynthia.)

Many things that were once supposed to be known concerning Venus are not confirmed by later and better observations. Venus is surrounded by an atmosphere so dense with clouds that it is conceded that her time of rotation and the inclination of her axis cannot be determined. She revealed one of the grandest secrets of the universe to the first seeker; showed her highest beauty to her first ardent lover, and has veiled herself from the prying eyes of later comers.

Florence has built a kind of shrine for the telescope of Galileo. By it he discovered the phases of Venus, the spots on the sun, the mountains of the moon, the satellites of Jupiter, and some irregularities of shape in Saturn, caused by its rings. Galileo subsequently became blind, but he had used his eyes to the best purpose of any man in his generation.

THE EARTH.

Its sign ⊕.

Distance from the sun, 92,500,000 miles. Diameter, polar, 7899 miles; equatorial, 7925½ miles. Axial revolution, 23h. 56m. 4.09s.; orbital, 365.26. Orbital velocity per minute, 1102.8 miles.

Let us lift ourselves up a thousand miles from the earth. We see it as a ball hung upon nothing in empty space. As the drop of falling water gathers itself

into a sphere by its own inherent attraction, so the
earth gathers itself into a ball.   Noticing closely, we

Fig. 54.—Earth and Moon in Space.

see forms of continents outlined in bright relief, and
oceanic forms in darker surfaces.   We see that its axis
of revolution is nearly perpendicular to the line of light
from the sun.   One-half is always dark.   The sunrise
greets a new thousand miles every hour; the glories of

the sunset follow over an equal space, 180° behind. We are glad that the darkness never overtakes the morning.

The Aurora Borealis.

While east and west are gorgeous with sunrise and sunset, the north is often more glorious with its aurora borealis. We remember that all worlds have weird

Fig. 55.—The Aurora as Waving Curtains.

and inexplicable appendages. They are not limited to their solid surfaces or their circumambient air. The sun has its fiery flames, corona, zodiacal light, and perhaps a finer kind of atmosphere than we know. The earth is not without its inexplicable surroundings. It has not only its gorgeous eastern sunrise, its glorious western sunset, high above its surface in the clouds, but it also has its more glorious northern dawn far above its clouds and air. The realm of this royal splendor is as yet an unconquered world waiting for its Alexander. There are certain observable facts, viz., it prevails mostly near the arctic circle rather than the pole; it takes on various forms—cloud-like, arched, straight; it streams like banners, waves like curtains in the wind, is inconstant; is either the cause or result of electric disturbance; it is often from four hundred to six hundred miles above the earth, while our air cannot be over one hundred miles. It almost seems like a revelation to human eyes of those vast, changeable, panoramic pictures by which the inhabitants of heaven are taught.

Investigation has discovered far more mysteries than it has explained. It is possible that the same cause that produces sun-spots produces aurora in all space, visible in all worlds. If so, we shall see more abundant auroras at the next maximum of sun-spot, between 1880–84.

The Delicate Balance of Forces.

A soap-bubble in the wind could hardly be more flexible in form and sensitive to influence than is the earth. On the morning of May 9th, 1876, the earth's crust at Peru gave a few great throbs upward, by the action of expansive gases within. The sea fled, and returned in great waves as the land rose and fell. Then these waves fled away over the great mobile surface, and in less than five hours they had covered a space equal to half of Europe. The waves ran out to the Sandwich Islands, six thousand miles, at the rate of five hundred miles an hour, and arrived there thirty feet high. They not only sped on in straight radial lines, but, having run up the coast to California, were deflected away into the former series of waves, making the most complex undulations. Similar beats of the great heart of the earth have sent its pulses as widely and rapidly on previous occasions.

The figure of the earth, even on the ocean, is irregular, in consequence of the greater preponderance of land — and hence greater density — in the northern hemisphere. These irregularities are often very perplexing in making exact geodetic measurements. The tendency of matter to fly from the centre by reason of revolution causes the equatorial diameter to be twenty-six miles longer than the polar one. By this force the Mississippi River is enabled to run up a hill nearly three miles high at a very rapid rate. Its mouth is that distance farther from the centre of the earth than its source, when but for this rotation both points would be equally distant.

If the water became more dense, or if the world were to revolve faster, the oceans would rush to the equator, burying the tallest mountains and leaving polar regions bare. If the water should become lighter in a very slight degree, or the world rotate more slowly, the poles would be submerged and the equator become an arid waste. No balance, turning to $\frac{1}{1000}$ of a grain, is more delicate than the poise of forces on the world. Laplace has given us proof that the period of the earth's axial rotation has not changed $\frac{1}{100}$ of a second of time in two thousand years.

7

Tides.

But there is an outside influence that is constantly acting upon the earth, and to which it constantly responds. Two hundred and forty thousand miles from the earth is the moon, having $\frac{1}{81}$ the mass of the world. Its attractive influence on the earth causes the movable and nearer portions to hurry away from the more stable and distant, and heap themselves up on that part of the earth nearest the moon. Gravitation is inversely as the square of the distance; hence the water on the surface of the earth is attracted more than the body of the earth, some parts of which are eight thousand miles farther off; hence the water rises on the side next the moon. But the earth, as a whole, is nearer the moon than the water on the opposite side, and being drawn more strongly, is taken away from the water, leaving it heaped up also on the side opposite to the moon.

A subsidiary cause of tides is found in the revolution of the earth and moon about their common centre of gravity. Revolution about an axis through the centre of a sphere enlarges the equator by centrifugal force. Revolution about an axis touching the surface of a flexible globe converts it into an egg-shaped body, with the longer axis perpendicular to the axis of revolution. In Fig. 56 the point of revolution is seen at the centre of gravity at G; hence, in the revolution of earth and moon as one, a strong centrifugal force is caused at D, and a less one at C. This gives greater height to the tides than the attraction of the moon alone could produce.

If the earth had no axial revolution, the attractive
point where the tide rises would be carried around the

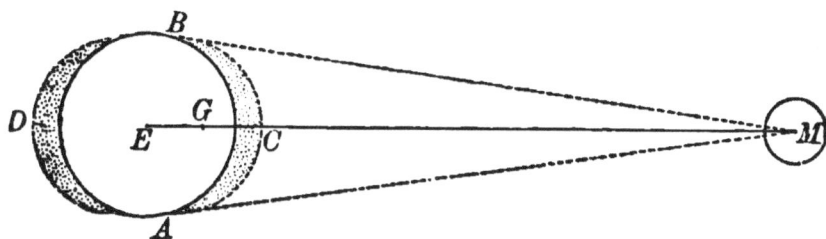

Fig. 56.

earth once in twenty-seven days by the moon's revolution about the earth. But since the earth revolves on
its axis, it presents a new section to the moon's attraction every hour. If the moon were stationary, that
would bring two high tides in exactly twenty-four
hours; but as the moon goes forward, we need nearly
twenty-five hours for two tides.

The attractive influence of the sun also gives us a
tide four-tenths as great as that of the moon. When
these two influences of the sun and moon combine, as
they do, in conjunction—when both bodies are on one
side of the earth; or in opposition, sun and moon being
on opposite sides of the earth—we have spring or increased tides. When the moon closes its first or third
quarter, *i. e.*, when a line from the moon to the earth
makes a right angle with one from the sun to the earth,
these influences antagonize one another, and we have
the neap or low tides.

It is easy to see that if, when the moon was drawing
its usual tide, the sun drew four-tenths of the water in
a tide at right angles with it, the moon's tide must be
by so much lower. Because of the inertia of the water

it does not yield instantly to the moon's influence, and the crest of the tide is some hours behind the advancing moon.

The amount of tide in various places is affected by almost innumerable influences, as distance of moon at its apogee or perigee; its position north, south, or at the equator; distance of earth from sun at perihelion and aphelion; the position of islands; the trend of continents, etc. All eastern shores have far greater tides than western. As the earth rolls to the east it leaves the tide-crest under the moon to impinge on eastern shores, hence the tides of from seventy-five to one hundred feet in the Bay of Fundy. Lakes and most seas are too small to have perceptible tides. The spring-tides in the Mediterranean Sea are only about three inches.

This constant ebb and flow of the great sea is a grand provision for its purification. Even the wind is sent to the sea to be cleansed. The sea washes every shore, purifies every cove, bay, and river twice every twenty-four hours. All putrescible matter liable to breed a pestilence is carried far from shore and sunk under fathoms of the never-stagnant sea. The distant moon lends its mighty power to carry the burdens of commerce. She takes all the loads that can be floated on her flowing tides, and cheerfully carries them in opposite directions in successive journeys.

It must be conceded that the profoundest study has not mastered the whole philosophy of tides. There are certain facts which are apparent, but for an explanation of their true theory such men as Laplace, Newton, and Airy have labored in vain. There are plenty of other worlds still to conquer.

Fig. 57.—Lunar Day.

THE MOON.

New moon, ● ; first quarter, ☽ ; full moon, ○ ; last quarter, ☾.

Extreme distance from the earth, 259,600 miles; least, 221,000 miles; mean, 240,000 miles. Diameter, 2164.6 miles [2153, Lockyer]. Revolution about the earth, 29½ days. Axial revolution, same time.

When the astronomer Herschel was observing the southern sky from the Cape of Good Hope, the most clever hoax was perpetrated that ever was palmed upon a credulous public. Some new and wonderful instruments were carefully described as having been used by that astronomer, whereby he was enabled to bring the moon so close that he could see thereon trees, houses, animals, and men-like human beings. He could even discern their movements, and gestures that indicated a peaceful race. The extent of the hoax will be perceived when it is stated that no telescope that we are now able to make reveals the moon more clearly than it would appear to the naked eye if it was one hundred or one hundred and fifty miles away. The distance at which a man can be seen by the unaided eye varies according to circumstances of position, background, light, and eye, but it is much inside of five miles.

Since, however, the moon is our nearest neighbor, a member of our own family in fact, it is a most interesting object of study.

A glance at its familiar face reveals its unequal illumination. All ages and races have seen a man in the moon. All lovers have sworn by its constancy, and only part of them have kept their oaths. Every twenty-nine or thirty days we see a silver crescent in the west, and are glad if it comes over the right shoulder—so much tribute does habit pay to superstition. The next night it is thirteen degrees farther east from the sun. We note the stars it occults, or passes by, and leaves behind as it broadens its disk, till it rises full-orbed in the east when the sun sinks in the west. It is easy to see that the moon goes around the earth from west to east. Afterward it rises later and smaller each night, till at length, lost from sight, it rises about the same time as the sun, and soon becomes the welcome crescent new moon again.

The same peculiarities are always evident in the visible face of the moon; hence we know that it always presents the same side to the earth. Obviously it must make just one axial to one orbital revolution. Hold any body before you at arm's-length, revolve it one-quarter around you until exactly overhead. If it has not revolved on an axis between the hands, another quarter of the surface is visible; but if in going up it is turned a quarter over, by the hands holding it steady, the same side is visible. Three causes enable us to see a little more than half the moon's surface: 1. The speed with which it traverses the ellipse of its orbit is variable. It sometimes gets ahead of us, sometimes behind, and we see farther around the front or back part. 2. The axis is a little inclined to the plane of its orbit, and its orbit a little inclined to ours; hence we see a little over its north pole, and then again over the south pole. 3. The earth being larger, its inhabitants see a little more than half-way around a smaller body. These causes combined enable us to see $\frac{576}{1000}$ of the moon's surface. Our eyes will never see the other side of the moon. If, now, being solid. her axial revolution could

be increased enough to make one more revolution in
two or three years, that difference between her axial
and orbital revolution would give the future inhabitants
of the earth a view of the entire circumference of the
moon. Yet if the moon were once in a fluid state, or
had oceans on the surface, the enormous tide caused by
the earth would produce friction enough, as they moved
over the surface, to gradually retard the axial revolution
till the two tidal elevations remained fixed toward and
opposite the earth, and then the axial and orbital revolutions would correspond, as at present. In fact, we can
prove that the form of the moon is protuberant toward
the earth. Its centre of gravity is thirty-three miles beyond its centre of magnitude, which is the same in effect as if a mountain of that enormous height rose on
the earth side. Hence any fluid, as water or air, would
flow round to the other side.

The moon's day, caused by the sun's light, is $29\frac{1}{2}$
times as long as ours. The sun shines unintermittingly
for fifteen days, raising a temperature as fervid as boil·
ing water. Then darkness and frightful cold for the
same time succeed, except on that half where the earth
acts as a moon. The earth presents the same phases—
crescent, full, and gibbous—to the moon as the moon
does to us, and for the same causes. Lord Rosse has
been enabled, by his six-foot reflector, to measure the
difference of heat on the moon under the full blaze of
its noonday and midnight. He finds it to be no less
than five hundred degrees. People not enjoying extremes of temperature should shun a lunar residence.
The moon gives us only $\frac{1}{818000}$ as much light as the sun.
A sky full of moons would scarcely make daylight.

7*

There are no indications of air or water on the moon. When it occults a star it instantly shuts off the light

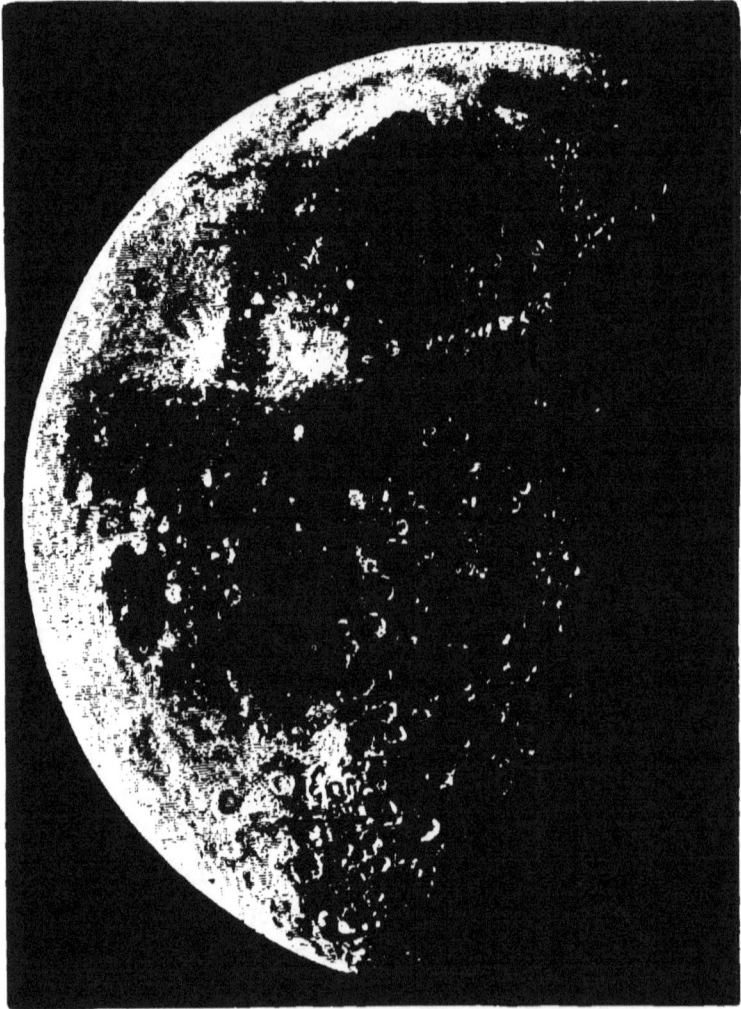

Fig. 58.—View of the Moon near the Third Quarter. From a Photograph by Professor Henry Draper.

and as instantly reveals it again. An atmosphere would gradually diminish and reveal the light, and by refrac-

tion cause the star to be hidden in much less time than
the solid body of the moon would need to pass over it.
If the moon ever had air and water, as it probably did,
they are now absorbed in the porous lava of its substance.

Telescopic Appearance.

Probably no one ever saw the moon by means of a
good telescope without a feeling of admiration and awe.
Except at full-moon, we can see where the daylight
struggles with the dark along the line of the moon's
sunrise or sunset. This line is called the terminator.
It is broken in the extreme, because the surface is as
rough as possible. In consequence of the small gravitation of the moon, utter absence of the expansive power
of ice shivering the cliffs, or the levelling power of rains,
precipices can stand in perpendicularity, mountains
shoot up like needles, and cav-
ities three miles deep remain
unfilled. The light of the
sun falling on the rough body
of the moon, shown in section (Fig. 59), illuminates the
whole cavity at *a*, part of the

Fig. 59.—Illumination of Craters and Peaks.

one at *b*, casts a long shadow from the mountain at *c*, and
touches the tip of the one at *d*, which appears to a distant
observer as a point of light beyond the terminator. As
the moon revolves the conical cavity, *a* is illuminated
on the forward side only; the light creeps down the
backward side of cavity *b* to the bottom; mountain *c*
comes directly under the sun and casts no shadow, and
mountain *d* casts its long shadow over the plain. Knowing the time of revolution, and observing the change of

illnmination, we can easily measure the height of mountain and depth of crater.   An apple, with excavations
and added prominences, revolved on its axis toward the

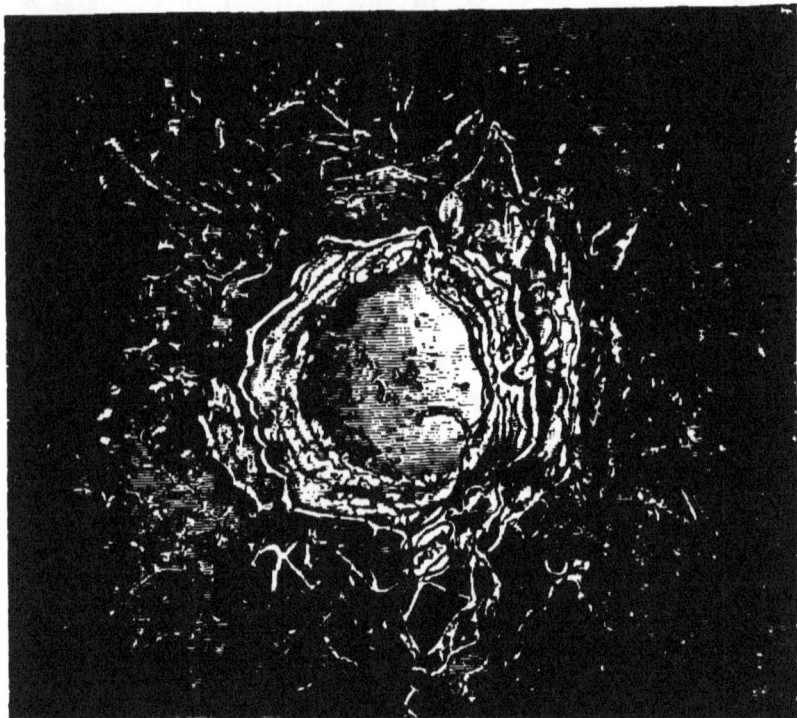

Fig. 60.—Lunar Crater " Copernicus," after Secchi.

light of a candle, admirably illustrates the crescent light
that fills either side of the cavities and the shadows of
the mountains on the plain.   Notice in Fig. 58 the crescent forms to the right, showing cavities in abnndance.

The selenography of one side of the moon is much
better known to us than the geography of the earth.
Our maps of the moon are far more perfect than those
of the earth ; and the photographs of lunar objects by
Messrs. Draper and De la Rue are wonderfully perfect,

and the drawings of Padre Secchi equally so (Fig. 60). The least change recognizable from the earth must be speedily detected. There are frequently reports of discoveries of volcanoes on the moon, but they prove to be illusions. The moon will probably look the same to observers a thousand years hence as it does to-day.

This little orb, that is only $\frac{1}{81}$ of the mass of the earth, has twenty-eight mountains that are higher than Mont Blanc, that "monarch of mountains," in Europe.

Eclipses.

It is evident that if the plane of the moon's orbit were to correspond with that of the earth, as they all lie in the plane of the page (Fig. 61), the moon must pass between the centres of the earth and sun, and exactly

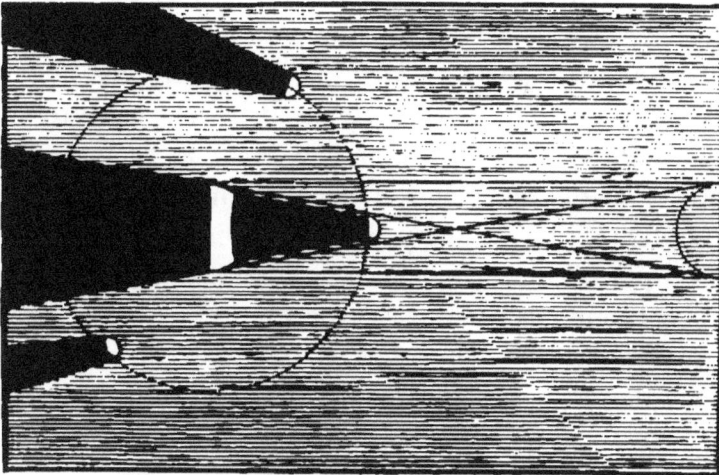

Fig. 61.—Eclipses; Shadows of Earth and Moon.

behind the earth at every revolution. Such successive and total darkenings would greatly derange all affairs dependent on light. It is easily avoided. Venus does

not cross the disk of the sun at every revolution, because of the inclination of the plane of its orbit to that
of the earth (see Fig. 41, p. 107). So the plane of the
orbit of the moon is inclined to the orbit of the earth
5° 8′ 39″; hence the full-moon is often above or below
the earth's shadow, and the earth is below or above the
moon's shadow at new moon. It is as if the moon's orbit were pulled up one-quarter of an inch from the page
behind the earth, and depressed as much below it between the earth and the sun. The point where the orbit of the moon penetrates the plane of the ecliptic is
called a node. If a new moon occur when the line of
intersection of the planes of orbits points to the sun,
the sun must be eclipsed; if the full-moon occur, the
moon must be eclipsed. In any other position the sun
or moon will only be partially hidden, or no eclipse will
occur.

If the new moon be near the earth it will completely
obscure the sun. A dime covers it if held close to the
eye. It may be so far from the earth as to only partially hide the sun; and, if it cover the centre, leave a
ring of sunlight on every side. This is called an annular eclipse. Two such eclipses will occur this year (1879).
If the full-moon passes near the earth, or is at perigee,
it finds the cone of shadow cast by the earth larger, and
hence the eclipse is greater; if it is far from the earth,
or near apogee, the earth's shadow is smaller, and the
eclipse less, or is escaped altogether.

There is a certain periodicity in eclipses. Whenever
the sun, moon, and earth are in a line, as in the total
eclipse of July 29th, 1878, they will be in the same position after the earth has made about eighteen revolu-

tions, and the moon two hundred and thirty-five—that is, eighteen years after. This period, however, is disregarded by astronomers, and each eclipse calculated by itself to the accuracy of a second.

How terrible is the fear of ignorance and superstition when the sun or moon appear to be in the process of destruction! how delightful are the joys of knowledge when its prophesies in regard to the heavenly bodies are being fulfilled!

MARS.

The god of war; its sign ♂, spear and shield.

Mean distance from the sun, 141,000,000 miles. Diameter, 4211 miles. Revolution, axial, 24h. 37m. 22.7s.; orbital, 686.98 days. Velocity per minute, 899 miles. Satellites, two.

At intervals, on an average of two years one month and nineteen days, we find rising, as the sun goes down, the reddest star in the heavens. Its brightness is exceedingly variable; sometimes it scintillates, and sometimes it shines with a steady light. Its marked peculiarities demand a close study. We find it to be Mars, the fiery god of war. Its orbit is far from circular. At perihelion it is 128,000,000 miles from the sun, and at aphelion 154,000,000; hence its mean distance is about 141,000,000. So great a change in its distance from the sun partly accounts for the change in its brilliancy. Now, if Mars and the earth revolved in circular orbits, the one 141,000,000 miles from the sun, and the other 92,000,000, they would approach at opposition within 49,000,000 miles of each other, and at conjunction be 233,000,000 miles apart. But Mars at perihelion may be only 128,000,000 miles from the sun, and earth at aphelion may be 94,000,000 miles from the sun. They are, then, but 34,000,000 miles apart. This favorable opportunity occurs about once in sixteen years. At its last occurrence, in 1877, Mars introduced to us his two satellites, that had never before been seen by man. In consequence of this greatly varying distance, the apparent size of Mars differs very much (Fig. 62).

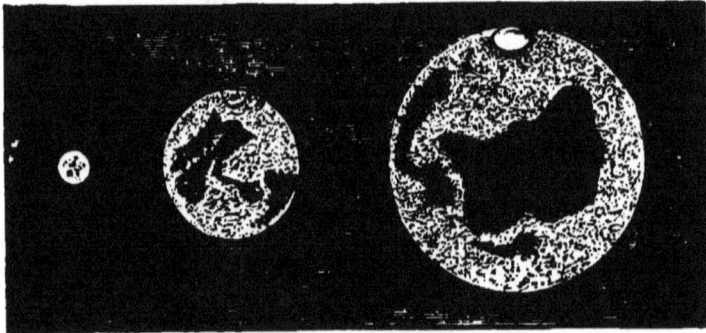

Fig. 62.—Apparent Size of Mars at Mean and Extreme Distances.

Take a favorable time when the planet is near, also as near overhead as it ever comes, so as to have as little atmosphere as possible to penetrate, and study the planet. The first thing that strikes the observer is a dazzling spot of white near the pole which happens to be toward him, or at both poles when the planet is so situated that they can be seen. When the north pole is turned toward the sun the size of the spot sensibly diminishes, and the spot at the south pole enlarges, and *vice versa.* Clearly they are ice-fields. Hence Mars has water, and air to carry it, and heat to melt ice. It is winter at the south pole when Mars is farthest from the sun; therefore the ice-fields are larger than at the north pole. It is summer at the south pole when Mars is nearest the sun. Hence its ice-fields grow smaller than those of the north pole in its summer. This carrying of water from pole to pole, and melting of ice over such large areas, might give rise to uncomfortable currents in ocean and air; but very likely an inhabitant of earth might be transported to the surface of Mars, and be no more surprised at what he observed there than if he went to some point of the earth to him unknown. Day and night would be nearly of the same length; winter would linger longer in the lap of spring; summer would be one hundred and eighty-one days long; but as the seas are more intermingled with the land, and the divisions of land have less of continental magnitude, it may be conjectured that Mars might be a comfortable place of residence to beings like men. Perhaps the greatest surprise to the earthly visitor would be to find himself weighing only four-tenths as much as usual, able to leap twice as high, and lift considerable bowlders.

Satellites of Mars.

The night of August 11th, 1877, is famous in modern astronomy. Mars has been a special object of study in all ages; but on that evening Professor Hall, of Washington, discovered a satellite of Mars. On the 16th it was seen again, and its orbital motion followed. On the following night it was hidden behind the body of the planet when the observation began, but at the calculated time—at four o'clock in the morning—it emerged, and established its character as a true moon, and not a fixed star or asteroid. Blessings, however, never come singly, for another object soon emerged which proved to be an inner satellite. This is extraordinarily near

the planet—only four thousand miles from the surface
—and its revolution is exceedingly rapid.　The shortest
period hitherto known is that of the inner satellite of
Saturn, 22h. 37m.　The inner satellite of Mars makes
its revolution in 7h. 39m.—a rapidity so much surpass-
ing the axial revolution of the planet itself, that it rises
in the west and sets in the east, showing all phases of
our moon in one night.　The outer satellite is 12,572
miles from Mars, and makes its revolution in 30h. 18m.
Its diameter is six and a quarter miles; that of the inner one is seven and a half miles.　This can be estimated only by the amount of light given.

ASTEROIDS.

**Already discovered (1879), 201.　Distances from the sun,
from 200,000,000 to 315,000,000 miles.　Diameters, from 20
to 400 miles.　Mass of all, less than one-quarter of the earth.**

The sense of infinite variety among the countless
number of celestial orbs has been growing rapidly upon
us for half a century, and doubtless will grow much
more in half a century to come.　Just as we paused in
the consideration of planets to consider meteors and
comets, at first thought so different, so must we now
pause to consider a ring of bodies, some of which are
as small in comparison to Jupiter, the next planet, as
aerolites are compared to the earth.

In 1800 an association of astronomers, suspecting that
a planet might be found in the great distance between
Mars and Jupiter, divided the zodiac into twenty-four
parts, and assigned one part to each astronomer for
a thorough search; but, before their organization could
commence work, Piazzi, an Italian astronomer of Paler-

mo, found in Taurus a star behaving like a planet. In six weeks it was lost in the rays of the sun. It was rediscovered on its emergence, and named Ceres. In March, 1802, a second planet was discovered by Olbers in the same gap between Mars and Jupiter, and named Pallas. Here was an embarrassment of richness. Olbers suggested that an original planet had exploded, and that more pieces could be found. More were found, but the theory is exploded into more pieces than a planet could possibly be. Up to 1879 one hundred and ninety-two have been discovered, with a prospect of more. Between 1871–75 forty-five were discovered, showing that they are sought for with great skill. In the discovery of these bodies, our American astronomers, Professors Watson and Peters, are without peers.

Between Mars and Jupiter is a distance of some 339,000,000 miles. Subtract 35,000,000 miles next to Mars and 50,000,000 miles next to Jupiter, and there is left a zone 254,000,000 miles wide outside of which the asteroids never wander. If any ever did, the attraction of Mars or Jupiter may have prevented their return.

Since the orbits of Mars and Jupiter show no sign of being affected by these bodies for a century past, it is probable that their number is limited, or at least that their combined mass does not approximate the size of a planet. Professor Newcomb estimates that if all that are now discovered were put into one planet, it would not be over four hundred miles in diameter; and if a thousand more should exist, of the average size of those discovered since 1850, their addition would not increase the diameter to more than five hundred miles.

That all these bodies, which differ from each other in no respect except in brilliancy, can be noted and fixed so as not to be mistaken one for another, and instantly recognized though not seen for a dozen years, is one of the highest exemplifications of the accuracy of astronomical observation.

JUPITER.

The king of the gods; sign ♃, the bird of Jove.

Distance from the sun, perihelion, 457,000,000 miles; aphelion, 503,000,000 miles. Diameter, equatorial, 87,500 miles; polar, 82,500 miles. Volume, 1400 earths. Mass, 213 earths. Axial revolution, 9h. 55m. 20s. Orbital revolution, 11 years 317 days. Velocity, 483.6 miles per minute.

Jupiter rightly wears the name of the " giant planet." His orbit is more nearly circular than most smaller

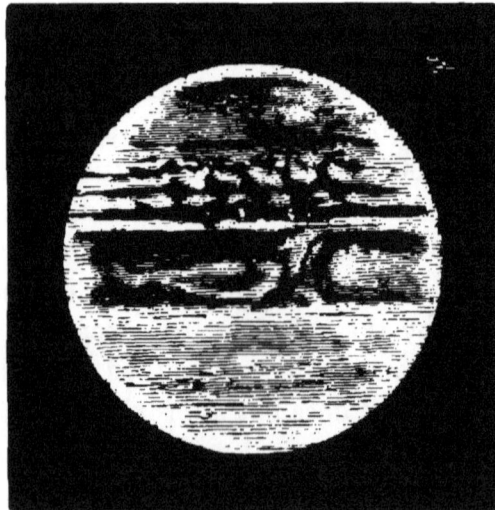

Fig. 63.—Jupiter as seen by the great Washington Telescope. Drawn by Mr. Holden.

planets. He could not turn short corners with facility. We know little of his surface. His spots and belts are changeable as clouds, which they probably are. Some spots may be slightly self-luminous, but not the part of the planet we see. It is covered with an enormous depth of atmosphere. Since the markings in the belts move about one hundred miles a day, the Jovian tempests are probably not violent. It is, however, a singular and unaccountable fact, as remarked by Arago, that its trade-winds move in an opposite direction from ours. Jupiter receives only one twenty-seventh as much light and heat from the sun as the earth receives. Its lighter density, being about that of water, indicates that it still has internal heat of its own. Indeed, it is likely that this planet has not yet cooled so as to have any solid crust, and if its dense vapors could be deposited on the surface, its appearance might be more suggestive of the sun than of the earth. (See note on recent red spot, p. 191.)

Satellites of Jupiter.

In one respect Jupiter seems like a minor sun—he is royally attended by a group of planets: we call them moons. This system is a favorite object of study to every one possessing a telescope. Indeed, I have known a man who could see these moons with the naked eye, and give their various positions without mistake. Galileo first revealed them to ordinary men. We see their orbits so nearly on the edge that the moons seem to be sliding back and forth across and behind the disk, and to varying distances on either side. Fig. 64 is the representation of their appearance at successive observations in November, 1878. Their motion is so swift, and the means of comparison by one another and the planet so excellent, that they can be seen to change their places,

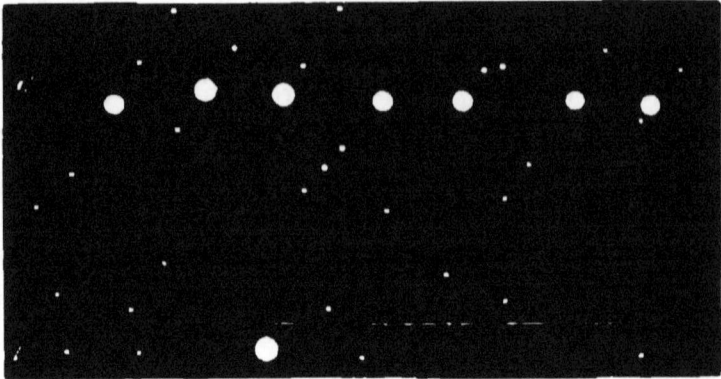

Fig. 64.—*a.* Various Positions of Jupiter's Moons; *b.* Greatest Elongation of each Satellite.

be occulted, emerge from shadow, and eclipse the planet, in an hour's watching.

ELEMENTS OF JUPITER'S SATELLITES.

	Mean Distance from Jupiter.	Sidereal Period.			Diameter.
	Miles.	Days	Hrs.	Min.	Miles.
I. Io...................	260,000	1	18	28	2,352
II. Europa...........	414,000	3	13	43	2,099
III. Ganymede......	661,000	7	3	59	3,436
IV. Callisto..........	1,162,000	16	18	5	2,929

It is seen by the above table that all these moons
are larger than ours, one larger than Mercury, and the
asteroids are hardly large enough to make respectable
moons for them. They differ in color: I. and II. have
a bluish tinge; III. a yellow; and IV. is red. The
amount of light given by these satellites varies in the
most sudden and inexplicable manner. Perhaps it may
be owing to the different distributions of land and water on them. The mass of all of them is .000171 of
Jupiter.

If the Jovian system were the only one in existence, it would be a surprising object of wonder and study. A monster planet, 85,000 miles in diameter, hung on nothing, revolving its equatorial surface over 450 miles a minute, holding four other worlds in steady orbits, some of them at a speed of 700 miles a minute, and the whole system carried through space at 500 miles a minute. Yet the discovery of all this display of power, skill, and stability is only reading the easiest syllables of the vast literature of wisdom and power.

SATURN.

The god of time; sign ♄, his scythe.

Mean distance from the sun, 881,000,000 miles. Diameter, polar, 66,500 miles; equatorial, 73,300 miles. Axial revolution, 10h. 14m. Periodic time, 29½ years. Moons, eight.

The human mind has used Saturn and the two known planets beyond for the last 200 years as a gymnasium. It has exercised itself in comprehending their enormous distances in order to clear those greater spaces, to where the stars are set; it has exercised its ingenuity at interpreting appearances which signify something other than they seem, in order that it may no longer be deluded by any sunrises into a belief that the heavenly dome goes round the earth. That a wandering point of light should develop into such amazing grandeurs under the telescope, is as unexpected as that every tiny seed should show peculiar markings and colors under the microscope.

There are certain things that are easy to determine, such as size, density, periodic time, velocity, etc.; but other things are exceedingly difficult to determine. It requires long sight to read when the book is held

800,000,000 miles away. Only very few, if more than two, opportunities have been found to determine the time of Saturn's rotation. On the evening of December

Fig. 65.—View of Saturn and his Rings.

7th, 1870, Professor Hall observed a brilliant white spot suddenly show itself on the body of this planet. It was as if an eruption of white hot matter burst up from the interior. It spread eastward, and remained bright till January, when it faded. No such opportunity for getting a basis on which to found a calculation of the time of the rotation of Saturn has occurred since Sir William Herschel's observations; and, very singularly, the two times deduced wonderfully coincide—that of Herschel being 10h. 16m., that of Mr. Hall being 10h. 14m.

The density of Saturn is less than that of water, and its velocity of rotation so great that centrifugal force antagonizes gravitation to such an extent that bodies weigh on it about the same as on the earth. All the fine fancies of the habitability of this vaporous world, all the calculations of the number of people that could live on the square miles of the planet and its enormous rings, are only fancy. Nothing could live there with more brains than a fish, at most. It is a world in formative processes. We cannot hear the voice of the Creator there, but we can see matter responsive to the voice, and moulded by his word.

Rings of Saturn.

The eye and mind of man have worked out a problem of marvellous difficulty in finding a true solution of the strange appearance of the rings. Galileo has the immortal honor of first having seen something peculiar about this planet. He wrote to the Duke of Tuscany, " When I view Saturn it seems *tricorps*. The central body seems the largest. The two others, situated, the one on the east, and the other on the west, seem to touch it. They are like two supporters, who help old Saturn on his way, and always remain at his side." Looking a few years later, the rings having turned from view, he said, " It is possible that some demon mocked me ;" and he refused to look any more.

Huyghens, in March, 1655, solved the problem of the triform appearance of Saturn. He saw them as handles on the two sides. In a year they had disappeared, and the planet was as round as it seemed to Galileo in 1612. He did not, however, despair; and in October,

1656, he was rewarded by seeing them appear again. He wrote of Saturn, " It is girdled by a thin plain ring, nowhere touching, inclined to the ecliptic."

Since that time discoveries have succeeded one another rapidly. " We have seen by degrees a ring evolved out of a triform planet, and the great division of the ring and the irregularities on it brought to light. Enceladus, and coy Mimas, faintest of twinklers, are caught by Herschel's giant mirrors. And he, too, first of men, realizes the wonderful tenuity of the ring, along which he saw those satellites travelling like pearls strung on a silver thread. Then Bond comes on the field, and furnishes evidence to show that we must multiply the number of separate rings we know not how many fold. And here we reach the golden age of Saturnian discovery, when Bond, with the giant refractor of Cambridge, and Dawes, with his 6½-inch Munich glass, first beheld that wonderful dark semi-transparent ring, which still remains one of the wonders of our system. But the end is not yet: on the southern surface of the ring, ere summer fades into autumn, Otto Struve in turn comes upon the field, detects, as Dawes had previously done, a division even in the dark ring, and measures it, while it is invisible to Lassell's mirror—a proof, if one were needed, of the enormous superiority possessed by refractors in such inquiries. Then we approach 1861, when the ring plane again passes through the earth, and Struve and Wray observe curious nebulous appearances."*

Our opportunities for seeing Saturn vary greatly. As the earth at one part of its orbit presents its south pole

---

* Lockyer.

to the sun, then its equator, then the north pole, so Saturn; and we, in the direction of the sun, see the south side of the rings inclined at an angle of 27°; next the edge of the rings, like a fine thread of light; then the north side at a similar inclination. On February 7th, 1878, Saturn was between Aquarius and Pisces, with the edge of the ring to the sun. In 1885, the planet being in Taurus, the south side of the rings will be seen at the greatest advantage. From 1881 till 1885 all circumstances will combine to give most favorable studies of Saturn. Meanwhile study the picture of it. The outer ring is narrow, dark, showing hints of another division, sometimes more evident than at others, as if it were in a state of flux. The inner, or second, ring is much brighter, especially on the outer edge, and shading off to the dusky edge next to the planet. There is no sign of division into a third dusky innermost ring, as was plainly seen by Bond. This, too, may be in a state of flux.

The markings of the planet are delicate, difficult of detection, and are not like those stark zebra stripes that are so often represented.

The distance between the planet and the second ring seems to be diminished one-half since 1657, and this ring has doubled its breadth in the same time. Some of this difference may be owing to our greater telescopic power, enabling us to see the ring closer to the planet; but in all probability the ring is closing in upon the central body, and will touch it by A.D. 2150. Thus the whole ring must ultimately fall upon the planet, instead of making a satellite.

We are anxious to learn the nature of such a ring.

Laplace mathematically demonstrated that it cannot be uniform and solid, and survive. Professor Peirce showed it could not be fluid, and continue. Then Professor Maxwell showed that it must be formed of clouds of satellites too small to be seen individually, and too near together for the spaces to be discerned, unless, perhaps, we may except the inner dark ring, where they are not near enough to make it positively luminous. Indeed, there is some evidence that the meteoroids are far enough apart to make the ring partially transparent.

We look forward to the opportunities for observation in 1882 with the brightest hope that these difficult questions will be solved.

Satellites of Saturn.

The first discovered satellite of Saturn seen by Huyghens was in 1655, and the last by the Bonds, father and son, of Cambridge, in 1848. These are eight in number, and are named:

		Distant from Saturn's centre.
I.	Mimas	119,725 miles.
II.	Enceladus	153,630 "
III.	Tethys	190,225 "
IV.	Dione	243,670 "
V.	Rhea	340,320 "
VI.	Titan	788,915 "
VII.	Hyperion	954,160 "
VIII.	Japetus	2,292,790 "

Titan can be seen by almost any telescope; I., II., and VII., only by the most powerful instrument. All except Japetus revolve nearly in the plane of the ring. Like the moons of Jupiter, they present remarkable and unaccountable variations of brilliancy. An inspection of the table reveals either an expectation that another moon will be discovered between V. and VI., and about three more between VII. and VIII., or that these gaps may be filled with groups of invisible asteroids, as the gap between Mars and Jupiter. This will become more evident by drawing Saturn, the rings, and orbits of the moons all as circles, on a scale of 10,000 miles to the inch. Saturn will be in the centre, 70,000 miles in diameter; then a gap, decreasing twenty-nine miles a year to the first ring, of, say, 10,000 miles; a dark ring 9000 miles wide; next the brightest ring 18,300 miles wide; then a gap of 1750 miles; then the outer ring 10,000 miles wide; then the orbits of the satellites in order.

If the scenery of Jupiter is magnificent, that of Saturn must be sublime. If one could exist there, he might wander from the illuminated side of the rings, under their magnificent arches, to the darkened side, see the swift whirling moons; one of them presenting ten times the disk of the earth's moon, and so very near as to enable him to watch the advancing line of light that marks the lunar morning journeying round that orb.

URANUS.

Sign ♅; the initial of Herschel, and sign of the world.

Distance from the sun, 1,771,000,000 miles. Diameter, 31,700 miles. Axial revolution unknown. Orbital, 84 years. Velocity per minute, 252 miles. Moons, four.

Uranus was presented to the knowledge of man as an unexpected reward for honest work. It was first mistaken by its discoverer for a comet, a mere cloud of vapor; but it proved to be a world, and extended the boundaries of our solar system, in the moment of its discovery, as much as all investigation had done in all previous ages.

Sir William Herschel was engaged in mapping stars in 1781, when he first observed its sea-green disk. He proposed to call it *Georgium Sidus*, in honor of his king; but there were too many names of the gods in the sky to allow a mortal name to be placed among them. It was therefore called Uranus, since, being the most distant body of our system, as was supposed, it might appropriately bear the name of the oldest god. Finding anything in God's realms of infinite riches ought not to lead men to regard that as final, but as a promise of more to follow.

This planet had been seen five times by Flamsteed before its character was determined—once nearly a century before—and eight times by Le Monnier. These names, which might easily have been associated with a grand discovery, are associated with careless observation. Eyes were made not only to be kept open, but to have minds behind them to interpret their visions. Herschel thought he discovered six moons belonging to Uranus, but subsequent investigation has limited the number to four. Two of these are seen with great difficulty by the most powerful telescopes.

If the plane of our moon's orbit were tipped up to a greater inclination, revolving it on the line of nodes as an axis until it was turned 85°, the moon, still continuing on its orbit in that plane, would go over the poles instead of about the equator, and would go back to its old path when the plane was revolved 180°; but its revolution would now be from east to west, or retrograde. The plane of the moons of Uranus has been thus inclined till it has passed 10° beyond the pole, and the moons' motions are retrograde as regards other known celestial movements. How Uranus itself revolves is not known. There are more worlds to conquer.

NEPTUNE.

God of the sea; sign ♆, his trident.

Distance from the sun, 2,775,000,000 miles. Diameter, 34,500 miles. Velocity per minute, 2016 miles. Axial revolution unknown. Orbital, 164.78 years. One moon.

Men sought for Neptune, as the heroes sought the golden fleece. The place of Uranus had been mapped for nearly one hundred years by these accidental observations. On applying the law of universal gravitation, a slight discrepancy was found between its computed place and its observed place. This discrepancy was exceedingly slight. In 1830 it was only 20''; in 1840, 90''; in 1844, 2'. Two stars that were 2' apart would appear as one to the keenest unaided eye, but such an error must not exist in astronomy. Years of work were given to its correction. Mr. John C. Adams, of Cambridge, England, finding that the attraction of a planet exterior to Uranus would account for its irregularities, computed the place of such a hypothetical body with singular exactness in October, 1841; but neither he nor the royal astronomer Airy looked for it. Another opportunity for immortality was heedlessly neglected. Meanwhile, M. Leverrier, of Paris, was working at the same problem. In the summer of 1846 Leverrier announced the place of the exterior planet. The conclusion was in striking coincidence with that of Mr.

Adams. Mr. Challis commenced to search for the planet near the indicated place, and actually saw and mapped the star August 4th, 1846, but did not recognize its planetary character. Dr. Galle, of Berlin, on the 23d of September, 1846, found an object with a planetary disk not plotted on the map of stars. It was the sought-for world. It would seem easy to find a world seventy-six times as large as the earth, and easy to recognize it when seen. The fact that it could be discovered only by such care conveys an overwhelming idea of the distance where it moves.

The effect of these perturbations by an exterior planet is understood from Fig. 66. Uranus and Neptune

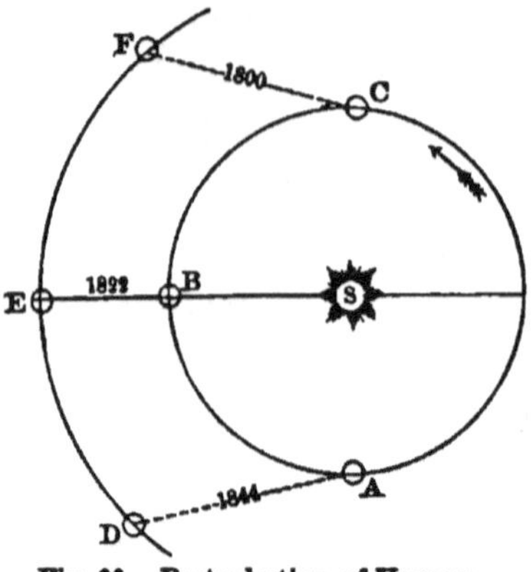

Fig. 66.—Perturbation of Uranus.

were in conjunction, as shown, in 1822. But in 1820 it had been found that Uranus was too far from the sun, and too much accelerated. Since 1800, Neptune, in his orbit from F to E, had been hastening Uranus in his orbit from C to B, and also drawing it farther from the sun. After 1822, Neptune, in passing from E to D, had been retarding Uranus in his orbit from B to A.

We have seen it is easy to miss immortality. There is still another instance. Lalande saw Neptune on May 8th and 10th, 1795, noted that it had moved a little, and that the observations did not agree; but, supposing the first was wrong, carelessly missed the glory of once more doubling the bounds of the empire of the sun.

It is time to pause and review our knowledge of this system. The first view reveals a moon and earth endowed with a force of inertia going on in space in straight lines; but an invisible elastic cord of attraction holds them together, just counterbalancing this tendency to fly apart, and hence they circle round their centre of gravity. The revolving earth turns every part of its surface to the moon in each twenty-five hours. By an axial revolution in the same time that the moon goes round the earth, the moon holds the same point of its surface constantly toward the earth. If we were to add one, two, four, eight moons at appropriate distances, the result would be the same. There is, however, another attractive influence—that of the sun. The sun attracts both earth and moon, but their nearer affection for each other keeps them from going apart. They both, revolving on their axes and around their centre of gravity, sweep in a vastly wider curve around the sun. Add as many moons as has Jupiter or Saturn, the result is the same—an orderly carrying of worlds through space.

There lies the unsupported sun in the centre, nearer to infinity in all its capacities and intensities of force than our minds can measure, filling the whole dome to where the stars are set with light, heat, and power. It holds four small worlds, namely, Mercury, Venus, Earth, and Mars—within a space whose radius it would require a locomotive half a thousand years to traverse. It next holds some indeterminate number of asteroids, and the great Jupiter, equal in volume to 1400 earths. It holds Saturn, Uranus, and Neptune, and all their variously related satellites and rings. The two thoughts that overwhelm us are distance and power. The period of man's whole history is not sufficient for an express train to traverse half the distance to Neptune. Thought wearies and fails in seeking to grasp such distances; it can scarcely comprehend one million miles, and here are thousands of them. Even the wings of imagination grow weary and droop. When we stand on that outermost of planets, the very last sentinel of the outposts of the king, the very sun grown dim and small in the distance, we have taken only one step of the infinite distance to the stars. They have not changed their relative position—they have not grown brighter by our approach. Neptune carries us round a vast circle about the centre of the dome of stars, but we seem no nearer its sides. In visiting planets, we have been only visiting next-door neighbors in the streets of a seaport town. We know that there are similar neighbors about Sirius and Arcturus, but a vast sea rolls between. As we said, we stand with the outermost sentinel; but into the great void beyond the king of day sends his comets as scouts, and they fly thousands of years without for one instant missing the steady grasp of the power of the sun. It is nearer almightiness than we are able to think.

If we cannot solve the problems of the present existence of worlds, how little can we expect to fathom the unsoundable depths of their creation and development through ages measureless to man! Yet the very difficulty provokes the most ambitious thought. We toil at the problem because it has been hitherto unsolvable. Every error we make, and discover to be such, helps toward the final solution. Every earnest thinker who climbs the shining worlds as steps to a higher thought is trying to solve the problem God has given us to do.

IX.

THE NEBULAR HYPOTHESIS.

"And the earth was without form, and void; and darkness was upon the face of the deep."—*Genesis* i. 2.

> " A dark
> Illimitable ocean, without bound,
> Without dimension, where length, breadth, and height,
> And time, and place are lost."—MILTON.

"It is certain that matter is somehow directed, controlled, and arranged ; while no material forces or properties are known to be capable of discharging such functions."—LIONEL BEALE.

"The laws of nature do not account for their own origin."—JOHN STUART MILL.

IX.

THE NEBULAR HYPOTHESIS.

THE method by which the solar system came into its present form was sketched in vast outline by Moses. He gave us the fundamental idea of what is called the nebular hypothesis. Swedenborg, that prodigal dreamer of vagaries, in 1734 threw out some conjectures of the way in which the outlines were to be filled up; Buffon followed him closely in 1749; Kant sought to give it an ideal philosophical completeness, as he said, "not as the result of observation and computation," but as evolved out of his own consciousness; and Laplace sought to settle it on a mathematical basis.

It has been modified greatly by later writers, and must receive still greater modifications before it can be accepted by the best scientists of to-day. It has been called "the grandest generalization of the human mind;" and if it shall finally be so modified as to pass from a tentative hypothesis to an accepted philosophy, declaring the modes of a divine worker rather than the necessities of blind force, it will still be worthy of that high distinction.

Let it be clearly noted that it never proposes to do more than to trace a portion of the mode of working which brought the universe from one stage to another. It only goes back to a definite point, never to absolute beginning, nor to nothingness. It takes matter from the hand of the unseen power behind, and merely notes the progress of its development. It finds the clay in the hands of an intelligent potter, and sees it whirl in the process of formation into a vessel. It is not in any sense necessarily atheistic, any more than it is to affirm that a tree grows by vital processes in the sun and dew, instead of being arbitrarily and instantly created. The conclusion reached depends on the spirit of the observer. Newton could say, "This most beautiful system of the sun, planets, and comets could only proceed from the counsel and dominion of an intelligent and powerful being!" Still it is well to recognize that some of its most ardent defenders have advocated it as materialistic. And Laplace said of it to Napoleon, "I have no need of the hypothesis of a god."

The materialistic statement of the theory is this: that matter is at first assumed to exist as an infinite cloud of fire-mist, dowered with power latent therein to grow of itself into every possibility of world, flower, animal, man, mind, and affection, without any interference or help from without. But it requires far more of the Divine Worker than any other theory. He must fill matter with capabilities to take care of itself, and this would tax the abilities of the Infinite One far more than a constant supervision and occasional interference. Instead of making the vase in perfect form, and coloring it with exquisite beauty by an ever-present skill, he must endow the clay with power to make itself in perfect form, adorn itself with delicate beauty, and create other vases.

The nebular hypothesis is briefly this: All the matter composing all the bodies of the sun, planets, and satellites once existed in an exceedingly diffused state;

rarer than any gas with which we are acquainted, filling a space larger than the orbit of Neptune. Gravitation gradually contracted this matter into a condensing globe of immense extent. Some parts would naturally be denser than others, and in the course of contraction a rotary motion, it is affirmed, would be engendered. Rotation would flatten the globe somewhat in the line of its axis. Contracting still more, the rarer gases, aided by centrifugal force, would be left behind as a ring that would ultimately be separated, like Saturn's ring, from the retreating body. There would naturally be some places in this ring denser than others; these would gradually absorb all the ring into a planet, and still revolve about the central mass, and still rotate on its own axis, throwing off rings from itself. Thus the planet Neptune would be left behind in the first sun-ring, to make its one moon; the planet Uranus left in the next sun-ring, to make its four moons from four successive planet-rings; Saturn, with its eight moons and three rings not made into moons, is left in the third sun-ring; and so on down to Mercury.

The outer planets would cool off first, become inhabitable, and, as the sun contracted and they radiated their own heat, become refrigerated and left behind by the retreating sun. Of course the outer planets would move slowly; but as that portion of the sun which gave them their motion drew in toward the centre, keeping its absolute speed, and revolving in the lessening circles of a contracting body, it would give the faster motion necessary to be imparted to Earth, Venus, and Mercury.

The four great classes of facts confirmatory of this hypothesis are as follows: 1st. All the planets move in the same direction, and nearly in the same plane, as if thrown off from one equator; 2d. The motions of the satellites about their primaries are mostly in the same direction as that of their primaries about the sun; 3d. The rotation of most of these bodies on their axes, and also of the sun, is in the same direction as the motion of the planets about the sun; 4th. The orbits of the planets, excluding asteroids, and their satellites, have but a comparatively small eccentricity; 5th. Certain nebulæ are observable in the heavens which are not yet condensed into solids, but are still bright gas.

The materialistic evolutionist takes up the idea of a universe of material world-stuff without form, and void, but so endowed as to develop itself into orderly worlds, and adds to it this exceeding advance, that when soil, sun, and chemical laws found themselves properly related, a force in matter, latent for a million eons in the original cloud, comes forward, and dead matter becomes alive in the lowest order of vegetable life; there takes place, as Herbert Spencer says, "a change from an indefinite, incoherent homogeneity, into a definite, coherent heterogeneity, through continuous differentiation and integration." The dead becomes alive; matter passes from unconsciousness to consciousness; passes up from plant to animal, from animal to man; takes on power to think, reason, love, and adore. The theistic evolutionist may think that the same process is gone through, but that an ever-present and working God superintends, guides, and occasionally bestows a new endowment of power that successively gives life, consciousness, mental, affectional, and spiritual capacity.

Is this world-theory true? and if so, is either of the

evolution theories true, also? If the first evolution theory is true, the evolved man will hardly know which to adore most, the Being that could so endow matter, or the matter capable of such endowment.

There are some difficulties in the way of the acceptance of the nebular hypothesis that compel many of the most thorough scientists of the day to withhold their assent to its entirety. The latest, and one of the most competent writers on the subject, Professor Newcomb, who is a mathematical astronomer, and not an easy theorist, evolving the system of the universe from the depth of his own consciousness, says: "Should any one be sceptical as to the sufficiency of these laws to account for the present state of things, science can furnish no evidence strong enough to overthrow his doubts until the sun shall be found to be growing smaller by actual measurement, or the nebulæ be actually seen to condense into stars and systems." In one of the most elaborate defences of the theory, it is argued that the hypothesis explains why only one of the four planets nearest the sun can have a moon, and why there can be no planet inside of Mercury. The discovery of the two satellites to Mars makes it all the worse for these facts.

Some of the objections to the theory should be known by every thinker. Laplace must have the cloud "diffused in consequence of excessive heat," etc. Helmholtz, in order to account for the heat of the contracting sun, must have the cloud relatively cold. How he and his followers diffused the cloud without heat is not stated.

The next difficulty is that of rotation. The laws

of science compel a contraction into one non-rotating body—a central sun, indeed, but no planets about it. Laplace cleverly evades the difficulty by not taking from the hand of the Creator diffused gas, but a sun with an atmosphere filling space to the orbit of Neptune, and *already in revolution.* He says: "It is four millions to one that all motions of the planets, rotations and revolutions, were at once imparted by an original common cause, of which we know neither the nature nor the epoch." Helmholtz says of rotation, "the existence of which must be assumed." Professor Newcomb says that the planets would not be arranged as now, each one twice as far from the sun as the next interior one, and the outer ones made first, but that all would be made into planets at once, and the small inner ones quite likely to cool off more rapidly.

It is a very serious difficulty that at least one satellite does not revolve in the right direction. How Neptune or Uranus could throw their moons backward from its equator is not easily accounted for. It is at least one Parthian arrow at the system, not necessarily fatal, but certainly dangerous.

A greater difficulty is presented by the recently discovered satellites of Mars. The inner one goes round the planet in one-third part of the time of the latter's revolution. How Mars could impart three times the speed to a body flying off its surface that it has itself, has caused several defenders of the hypothesis to rush forward with explanations, but none with anything more than mere imaginary collisions with some comet. It is to be noticed that accounting for three times the speed is not enough; for as Mars shrunk away from the ring that formed that satellite, it ought itself to attain more speed, as the sun revolves faster than its planets, and the earth faster than its moon. In defending the hypothesis, Mitchel said: "Suppose we had discovered that it required more time for Saturn or Jupiter to rotate on their axes than for their nearest moon to revolve round them in its orbit; this would have falsified the theory." The defenders of the nebular hypothesis avowed that certain conditions must be fatal to its acceptance. Later discoveries have established these very conditions as incontrovertible facts.

In regard to one Martial moon, Professor Kirkwood, on whom Proctor conferred the highest title that could be conferred, "the modern Kepler," says: "Unless some explanation can be given, the short period of the inner satellite will be doubtless regarded as a conclusive argument against the nebular hypothesis." If gravitation be sufficient to account for the various motions of the heavenly bodies, we have a perplexing problem in the star known as 1830 Goombridge, now in the Hunting Dogs of Bootes. It is thought to have a speed of two hundred miles per second — a velocity that all the known matter in the universe could not give to the star by all its combined attraction. Neither could all that attraction stop the motion of the star, or bend it into an orbit. Its motion must be accounted for on some hypothesis other than the nebular.

The nebulæ which we are able to observe are not altogether confirmatory of the hypothesis under consideration. They have the most fantastic shapes, as if they had no relation to rotating suns in the formative stages. There are vast gaps in the middle, where they ought to be densest. Mr. Plumer, in the *Natural Science Re-*

view, says, in regard to the results of the spectroscopic revelations: "We are furnished with distinct proof that the gases so examined are not only of nearly equal density, but that they exist in a low state of *tension. This fact is fatal to the nebular theory."*

In the autumn of 1876 a star blazed out in Cygnus, which promised to throw a flood of light on the question of world-making. Its spectrum was like some of the fixed stars. It probably blazed out by condensation from some previously invisible nebula. But its brilliancy diminished swiftly, when it ought to have taken millions of years to cool. If the theory was true, it ought to have behaved very differently. It should have regularly condensed from gas to a solid sun by slow process. But, worst of all, after being a star awhile, it showed unmistakable proofs of turning into a cloud-mist—a star into a nebula, instead of *vice versa*. A possible explanation will be considered under variable stars.

Such are a few of the many difficulties in the way of accepting the nebular hypothesis, as at present explained, as being the true mode of development of the solar system. Doubtless it has come from a hot and diffused condition into its present state; but when such men as Proctor, Newcomb, and Kirkwood see difficulties that cannot be explained, contradictions that cannot be reconciled by the principles of this theory, surely lesser men are obliged to suspend judgment, and render the Scotch verdict of "not proven." Whatever truth there may be in the theory will survive, and be incorporated into the final solution of the problem; which solution will be a much grander generalization of the human mind than the nebular hypothesis.

Of some things we feel very sure: that matter was once without form and void, and darkness rested on the face of the mighty deeps; that, instead of chaos, we have now cosmos and beauty; and that there is some process by which matter has been brought from one state to the other. Whether, however, the nebular hypothesis lays down the road travelled to this transfiguration, we are not sure. Some of it seems like solid rock, and some like shifting quicksand. Doubtless there is a road from that chaos to this fair cosmos. The nebular hypothesis has surveyed, worked, and perfected many long reaches of this road, but the rivers are not bridged, the chasms not filled, nor the mountains tunnelled.

When men attempt to roll the hypothesis of evolution along the road of the nebular hypothesis of worlds, and even beyond to the production of vegetable and animal life, mind and affection, the gaps in the road become evident, and disastrous.

A soul that has reached an adoration for the Supreme Father cares not how he has made him. Doubtless the way God chose was the best. It is as agreeable to have been thought of and provided for in the beginning, to have had a myriad ages of care, and to have come from the highest existent life at last, as to have been made at once, by a single act, ont of dust. The one who is made is not to say to the Maker, "Why hast thou formed me in this or that manner?" We only wish the question answered in what manner we were really made.

Evolution, without constant superintendence and occasional new inspiration of power, finds some tremendous chasms in the road it travels. These must be spanned by the power of a present God or the airy imagination of man. Dr. McCosh has happily enumerated some of these tremendous gaps over which mere force cannot go. Given, then, matter with mechanical power only, what are the gaps between it and spirituality?

"1. Chemical action cannot be produced by mechanical power.

"2. Life, even in the lowest forms, cannot be produced from unorganized matter.

"3. Protoplasm can be produced only by living matter.

"4. Organized matter is made up of cells, and can be produced only by cells. Whence the first cell?

"5. A living being can be produced only from a seed or germ. Whence the first vegetable seed?

"6. An animal cannot be produced from a plant. Whence the first animal?

"7. Sensation cannot be produced in insentient matter.

"8. The genesis of a new species of plant or animal has never come under the cognizance of man, either in pre-human or post-human ages, either in pre-scientific or scientific times. Darwin acknowledges this, and says that, should a new species suddenly arise, we have no means of knowing that it is such.

"9. Consciousness—that is, a knowledge of self and its operations—cannot be produced out of mere matter or sensation.

"10. We have no knowledge of man being generated out of the lower animals.

"11. All human beings, even savages, are capable of forming certain high ideas, such as those of God and duty. The brute creatures cannot be made to entertain these by any training.

"With such tremendous gaps in the process, the theory which would derive all things out of matter by development is seen to be a very precarious one."

The truth, according to the best judgment to be formed in the present state of knowledge, would seem to be about this: The nebular hypothesis is correct in all the main facts on which it is based; but that neither the present forces of matter, nor any other forces conceivable to the mind of man, with which it can possibly be endowed, can account for all the facts already observed. There is a demand for a personal volition, for an exercise of intelligence, for the following of a divine plan that embraces a final perfection through various and changeful processes. The five great classes of facts that sustain the nebular hypothesis seem set before us to show the regular order of working. The several facts that will not, so far as at present known, accord with that plan, seem to be set before us to declare the presence of a divine will and power working his good pleasure according to the exigencies of time and place.

NOTE TO PAGE 165.—The great red spot visible for years on Jupiter is still discernible (1886). Its appearance was a mystery; its shifting in longitude, its variation of distinctness, are mysteries still. It has given tangible proof of the equatorial acceleration of the planet. For the bright spots near the equator made a circuit around the planet in five minutes less time than the great red spot that was forty degrees from the equator. In precisely the same way the spots near the sun's equator complete a revolution in less time than those nearer the poles. Here is another link connecting the central luminary more intimately with his lordly son, and including his developments within the bounds of solar mysteries. When we find out the reason why the equatorial sun spots move faster than the polar sun spots, then we shall learn why the Jovian bright spots moved faster than the great red spot. We shall probably be convinced at the same time that the regal planet is far more in the condition of the sun than his less massive and less richly endowed brethren.

Richter says that "an angel once took a man and stripped him of his flesh, and lifted him up into space to show him the glory of the universe. When the flesh was taken away the man ceased to be cowardly, and was ready to fly with the angel past galaxy after galaxy, and infinity after infinity, and so man and angel passed on, viewing the universe, until the sun was out of sight—until our solar system appeared as a speck of light against the black empyrean, and there was only darkness. And they looked onward, and in the infinities of light before, a speck of light appeared, and suddenly they were in the midst of rushing worlds. But they passed beyond that system, and beyond system after system, and infinity after infinity, until the human heart sank, and the man cried out: 'End is there none of the universe of God?' The angel strengthened the man by words of counsel and courage, and they flew on again until worlds left behind them were out of sight, and specks of light in advance were transformed, as they approached them, into rushing systems; they moved over architraves of eternities, over pillars of immensities, over architecture of galaxies, unspeakable in dimensions and duration, and the human heart sank again and called out: 'End is there none of the universe of God?' And all the stars echoed the question with amazement: 'End is there none of the universe of God?' And this echo found no answer. They moved on again past immensities of immensities, and eternities of eternities, until in the dizziness of uncounted galaxies the human heart sank for the last time, and called out: 'End is there none of the universe of God?' And again all the stars repeated the question, and the angel answered: 'End is there none of the universe of God. Lo, also, there is no beginning.'"

X.

THE OPEN PAGE OF THE HEAVENS.

THE Greeks set their mythological deities in the skies, and read the revolving pictures as a starry poem. Not that they were the first to set the blazonry of the stars as monuments of their thought; we read certain allusions to stars and asterisms as far back as the time of Job. And the Pleiades, Arcturus, and Orion are some of the names used by Him who "calleth all the stars by their names, in the greatness of his power." Homer and Hesiod, 750 B.C., allude to a few stars and groups. The Arabians very early speak of the Great Bear; but the Greeks completely nationalized the heavens. They colonized the earth widely, but the heavens completely; and nightly over them marched the grand procession of their apotheosized divinities. There Hercules perpetually wrought his mighty labors for the good of man; there flashed and faded the changeful star Algol, as an eye in the head of the snaky-haired Medusa; over them flew Pegasus, the winged horse of the poet, careering among the stars; there the ship Argo, which had explored all strange seas of earth, nightly sailed in the infinite realms of heaven; there Perseus perpetually killed the sea-monster by celestial aid, and perpetually won the chained Andromeda for his bride. Very evident was their recognition of divine help: equally evident was their assertion of human ability and dominion. They gathered the illimitable stars, and put uncountable suns into the shape of the Great Bear — the most colossal form of animal ferocity and strength — across whose broad forehead imagination grows weary in flying; but they did not fail to put behind him a representative of themselves, who forever drives him around a sky that never sets—a perpetual type that man's ambition and expectation correspond to that which has always been revealed as the divine.

The heavens signify much higher power and wisdom to us; we retain the old pictures and groupings for the convenience of finding individual stars. It is enough for the astronomer that we speak of a star as, right ascension, 1h. 17m., declination, 88° 42'. But for most people, if not all, it is better to call it Polaris. So we might speak of a lake in latitude 42° 40', longitude 79° 22', but it would be clearer to most persons to say Chautauqua. For exact location of a star, right ascension and declination must be given; but for general indication its name or place in a constellation is sufficiently exact. The heaven is rather indeterminably laid out in irregular tracts, and the mythological names are preserved. The brightest stars are then indicated in order by the letters of the Greek alphabet—Alpha (a), Beta (β), Gamma (γ), etc. Stars are numbered in the order of right ascension without regard to their brilliancy. Thus No. 50 Cygni is Alpha; R. A., 20h. 36m. 19s. No. 61 is, R. A., 21h. 0m. 10s. Thus any star unknown by name is easily found. An acquaintance with the names, peculiarities, and movements of the stars visible at different seasons of the year is an unceasing source of pleasure. It

is not vision alone that is gratified, for one fine enough
may hear the morning stars sing together, and understand
the speech that day uttereth unto day, and the knowledge that night showeth unto night. One never can be
alone if he is familiarly acquainted with the stars. He
rises early in the summer morning, that he may see his
winter friends; in winter, that he may gladden himself
with a sight of the summer stars. He hails their successive rising as he does the coming of his personal
friends from beyond the sea. On the wide ocean he is
commercing with the skies, his rapt soul sitting in his
eyes. Under the clear skies of the East he hears God's
voice speaking to him, as to Abraham, and saying,
" Look now toward the heavens, and tell the number of
the stars, if thou be able to number them."

A general acquaintance with the stars will be first attempted; a more particular knowledge afterward. Fig.
67 (page 201) is a map of the circumpolar region, which
is in full view every clear night. It revolves daily
round Polaris, its central point. Toward this star, the
two end stars of the Great Dipper ever point, and are
in consequence called "the Pointers." The map may
be held toward the northern sky in such a position as
the stars may happen to be. The Great Bear, or Dipper,
will be seen at nine o'clock in the evening above the
pole in April and May; west of the pole, the Pointers
downward, in July and August; close to the north horizon in October and November; and east of the pole the
Pointers highest, in January and February. The names
of such constantly visible stars should be familiar. In
order, from the end of the tail of the Great Bear, we
have Benetnasch *η*, Mizar *ζ*, Little Alcor close to it, Ali-

9*

oth ε, Megrez δ, at the junction, has been growing dim-
mer for a century, Phad γ, Dubhe and Merak. It is
best to get some facility at estimating distances in degrees. Dubhe and Merak, "the Pointers," are five degrees apart. Eighteen degrees forward of Dubhe is
the Bear's nose; and three pairs of stars, fifteen degrees
apart, show the position of the Bear's three feet. Follow "the Pointers" twenty-nine degrees from Dubhe,
and we come to the pole-star. This star is double, made
of two suns, both appearing as one to the naked eye.
It is a test of an excellent three-inch telescope to resolve
it into two. Three stars beside it make the curved-up
handle of the Little Dipper of Ursa Minor. Between
the two Bears, thirteen degrees from Megrez, and eleven
degrees from Mizar, are two stars in the tail of the
Dragon, which curves about to appropriate all the stars
not otherwise assigned. Follow a curve of fifteen stars,
doubling back to a quadrangle from five to three degrees on a side, and thirty-five degrees from the pole, for
his head. His tongue runs out to a star four degrees
in front. We shall find, hereafter, that the foot of Hercules stands on this head. This is the Dragon slain by
Cadmus, and whose teeth produced such a crop of san-
guinary men.

The star Thuban was once the pole-star. In the
year B.C. 2300 it was ten times nearer the pole than
Polaris is now. In the year A.D. 2100 the pole will
be within 30′ of Polaris; in A.D. 7500, it will be at
a of Cepheus; in A.D. 13,500, within 7° of Vega; in
A.D. 15,700, at the star in the tongue of Draco; in
A.D. 23,000, at Thuban; in A.D. 28,000, back to Polaris.
This indicates no change in the position of the dome

of stars, but a change in the direction of the axis of
the earth pointing to these various places as the cycles pass. As the earth goes round its orbit, the axis,
maintaining nearly the same direction, really points to
every part of a circle near the north star as large as the
earth's orbit, that is, 185,000,000 miles in diameter. But,
as already shown, that circle is too small to be discernible at our distance. The wide circle of the pole through
the ages is really made up of the interlaced curves of
the annual curves continued through 25,870 years.
The stem of the spinning top wavers, describes a circle,
and finally falls ; the axis of the spinning earth wavers,
describes a circle of nearly 26,000 years, and never
falls.

The star γ Draconis, also called Etanin, is famous in
modern astronomy, because observations on this star
led to the discovery of the *aberration of light.* If we
held a glass tube perpendicularly out of the window of
a car at rest, when the rain was falling straight down,
we could see the drops pass directly through. Put
the car in motion, and the drops would seem to start
toward us, and the top of the tube must be bent forward,
or the drops entering would strike on the backside of
the tube carried toward them. So our telescopes are
bent forward on the moving earth, to enable the entered
light to reach the eye-piece. Hence the star does not
appear just where it is. As the earth moves faster in
some parts of its orbit than others, this aberration is
sometimes greater than at others. It is fortunate that
light moves with a uniform velocity, or this difficult
problem would be still further complicated. The displacement of a star from this cause is about 20″.43.

On the side of Polaris, opposite to Ursa Major, is King Cepheus, made of a few dim stars in the form of the letter K. Near by is his brilliant wife Cassiopeia, sitting on her throne of state. They were the graceless parents who chained their daughter to a rock for the sea-monster to devour; but Perseus, swift with the winged sandals of Mercury, terrible with his avenging sword, and invincible with the severed head of Medusa, whose horrid aspect of snaky hair and scaly body turned to stone every beholder, rescues the maiden from chains, and leads her away by the bands of love. Nothing could be more poetical than the life of Perseus. When he went to destroy the dreadful Gorgon, Medusa, Pluto lent him his helmet, which would make him invisible at will; Minerva loaned her buckler, impenetrable, and polished like a mirror; Mercury gave him a dagger of diamonds, and his winged sandals, which would carry him through the air. Coming to the loathsome thing, he would not look upon her, lest he, too, be turned to stone; but, guided by the reflection in the buckler, smote off her head, carried it high over Libya, the dropping blood turning to serpents, which have infested those deserts ever since.

The human mind has always been ready to deify and throne in the skies the heroes that labor for others. Both Perseus and Hercules are divine by one parent, and human by the other. They go up and down the earth, giving deliverance to captives, and breaking every yoke. They also seek to purge away all evil; they slay dragons, gorgons, devouring monsters, cleanse the foul places of earth, and one of them so wrestles with death as to win a victim from his grasp. Finally, by

Fig. 67.—Circumpolar Constellations. Always visible. In this position.—January 20th, at 10 o'clock; February 4th, at 9 o'clock; and February 19th, at 8 o'clock.

an ascension in light, they go up to be in light forever. They are not ideally perfect. They right wrong by slaying wrong-doers, rather than by being crucified themselves; they are just murderers; but that only plucks the fruit from the tree of evil. They never attempted to infuse a holy life. They punished rather than regenerated. It must be confessed, also, that they were not sinless. But they were the best saviors the race could imagine, and are examples of that perpetual effort of the human mind to incarnate a Divine Helper who shall labor and die for the good of men.

Fig. 68.—Algol is on the Meridian, 51° South of Pole.—At 10 o'clock, December 7th; 9 o'clock, December 22d; 8 o'clock, January 5th.

Equatorial Constellations.

If we turn our backs on Polaris on the 10th of November, at 10 o'clock in the evening, and look directly overhead, we shall see the beautiful constellation of Andromeda. Together with the square of Pegasus, it makes another enormous dipper. The star *a* Alpheratz is in her face, the three at the left cross her breast. β and the two above mark the girdle of her loins, and γ is in the foot. Perseus is near enough for help; and Cetus, the sea-monster, is far enough away to do no harm. Below, and east of Andromeda, is the Ram of the golden fleece, recognizable by the three stars in an acute triangle. The brightest is called Arietis, or Hamel. East of this are the Pleiades, and the V-shaped Hyades in Taurus, or the Bull. The Pleiades rise about 9 o'clock on the evening of the 10th of September, and at 3 o'clock A.M. on June 10th.

Fig. 69.—Capella (45° from the Pole) and Rigel (100°) are on the Meridian at 8 o'clock February 7th, 9 o'clock January 22d, and at 10 o'clock January 7th.

Fig. 69 extends east and south of our last map. It is the most gorgeous section of our heavens. (See the Notes to the Frontispiece.) Note the triangle, 26° on a side, made by Betelguese, Sirius, and Procyon. A line from Procyon to Pollux leads quite near to Polaris. Orion is the mighty hunter. Under his feet is a hare, behind him are two dogs, and before him is the rushing bull. The curve of stars to the right of Bellatrix, γ, represents his shield of the Nemean lion's hide. The three stars of his belt make a measure 3° long; the upper one, Mintaka, is less than 30' south of the equinoctial. The ecliptic passes between Aldebaran and the Pleiades. Sirius rises about 9 o'clock P.M. on the 1st of December, and about 4 o'clock A.M. on the 16th of August. Procyon rises about half an hour earlier.

Fig. 70.—Regulus comes on the Meridian, 79° south from the Pole, at 10 o'clock March 23d, 9 o'clock April 8th, and at 8 o'clock April 23d.

Fig. 70 continues eastward. Note the sickle in the head and neck of the Lion. The star β is Denebola, in his tail. Arcturus appears by the word Bootes, at the edge of the map. These two stars make a triangle with Spica, about 35° on a side. The geometric head of Hydra is easily discernible east of Procyon. The star γ in the Virgin is double, with a period of 145 years. ζ is just above the equinoctial. There is a fine nebula two-thirds of the way from δ to η, and a little above the line connecting the two. Coma Berenices is a beautiful cluster of faint stars. Spica rises at 9 o'clock on the 10th of February, at 5 o'clock A.M. on the 6th of November.

Fig. 71.—Arcturus comes to the Meridian, 70° from the Pole, at 10 o'clock May 25th, 9 o'clock June 9th, and at 8 o'clock June 25th.

Fig. 71 represents the sky to the eastward and northward of the last. A line drawn from Polaris and Benetnasch comes east of Arcturus to the little triangle called his sons. Bootes drives the Great Bear round the pole. Arcturus and Denebola make a triangle with *a*, also called Cor Cœroli, in the Hunting Dogs. This triangle, and the one having the same base, with Spica for its apex, is called the "Diamond of the Virgin." Hercules appears head down—*a* in the face, β, γ, δ in his shoulders, π and η in the loins, τ in the knee, the foot being bent to the stars at the right. The Serpent's head, making an ×, is just at the right of the γ of Hercules, and the partial circle of the Northern Crown above. The head of Draco is seen at β on the left of the map. Arcturus rises at 9 o'clock about the 20th of February, and at 5 A.M. on the 22d of October; Regulus 3h. 35m. earlier.

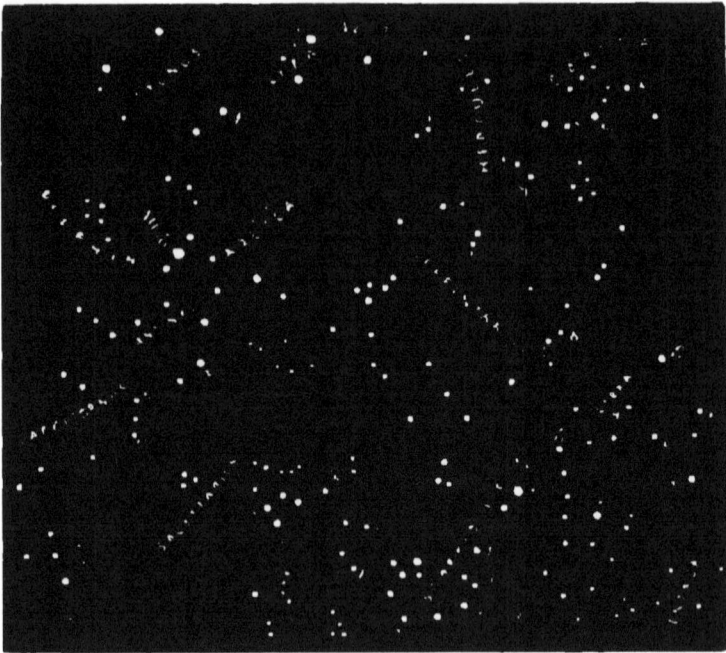

Fig. 72.—Altair comes to the Meridian, 82° from the Pole, at 10 o'clock P.M. August 18th, at 9 o'clock September 2d, and at 8 o'clock September 18th.

Fig. 72 portrays the stars eastward and southward. Scorpio is one of the most brilliant and easily traced constellations. Antares, *a*, in the heart, is double. In Sagittarius is the Little Milk-dipper, and west of it the bended bow. Vega is at the top of the map. Near it observe ζ, a double, and ε, a quadruple star. The point to which the solar system is tending is marked by the sign of the earth below π Herculis. The Serpent, west of Hercules, and coiled round nearly to Aquila, is very traceable. In the right-hand lower corner is the Centaur. Below, and always out of our sight, is the famous *a* Centauri. The diamond form of the Dolphin is sometimes called "Job's Coffin." The ecliptic passes close

Fig. 73.—Fomalhaut comes to the Meridian, only 17° from the horizon, at
8 o'clock November 4th.

to β of Scorpio, which star is in the head. Antares, in
Scorpio, rises at 9 o'clock P.M. on May 9th, and at 5
o'clock A.M. on January 5th.

In Fig. 73 we recognize the familiar stars of Pegasus,
which tell us we have gone quite round the heavens.
Note the beautiful cross in the Swan. β in the bill is
named Albireo, and is a beautiful double to almost any
glass. Its yellow and blue colors are very distinct.
The place of the famous double star 61 Cygni is seen.
The first magnitude star in the lower left-hand corner is
Fomalhaut, in the Southern Fish. a Pegasi is in the
diagonal corner from Alpheratz, in Andromeda. The
star below Altair is β Aquilæ, and is called Alschain;
the one above is γ Aquilæ, named Tarazed. This is
not a brilliant section of the sky. Altair rises at 9
o'clock on the 29th of May, and at 6 o'clock A.M. on
the 11th of January.

Fig. 74.—Southern Circumpolar Constellations invisible north of the Equator.

Fig. 74 gives the stars that are best seen by persons south of the earth's equator. In the ship is brilliant Canopus, and the remarkable variable η. Below it is the beautiful Southern Cross, near the pole of the southern heavens. Just below are the two first magnitude stars Bungula, α, and Achernar, β, of the Centaur. Such a number of unusually brilliant stars give the southern sky an unequalled splendor. In the midst of them, as if for contrast, is the dark hole, called by the sailors the "Coal-sack," where even the telescope reveals no sign of light. Here, also, are the two Magellanic clouds, both easily discernible by the naked eye; the larger two hundred times the apparent size of the moon, lying between the pole and Canopus, and the other between Achernar and the pole. The smaller cloud is only one-fourth the size of the other. Both are mostly resolvable into groups of stars from the fifth to the fifteenth magnitude.

For easy out-door finding of the stars above the horizon at any time, see star-maps at end of the book.

Characteristics of the Stars.

Such a superficial examination of stars as we have made scarcely touches the subject. It is as the study of the baptismal register, where the names were anciently recorded, without any knowledge of individuals. The heavens signify much more to us than to the Greeks. We revolve under a dome that investigation has infinitely enlarged from their estimate. Their little lights were turned by clumsy machinery, held together by material connections. Our vast worlds are connected by a force so fine that it seems to pass out of the realm of the material into that of the spiritual. Animal ferocity or a human Hercules could image their idea of power. Ours finds no symbol, but rises to the Almighty. Their heavens were full of fighting Orions, wild bulls, chained Andromedas, and devouring monsters. Our heavens are significant of harmony and unity; all worlds carried by one force, and all harmonized into perfect music. All their voices blend their various significations into a personal speaking, which says, "Hast thou not heard that the everlasting God, the Lord, the creator of the ends of the earth, fainteth not, neither is weary?" There is no searching of his understanding. Lift up your eyes on high, and behold who hath created all these things, that brought out their host by number, that calleth them all by their names in the greatness of his power; for that he is strong in power not one faileth.

Number.

We find about five thousand stars visible to the naked eye in the whole heavens, both north and south. Of these twenty are of the first magnitude, sixty-five of the second, two hundred of the third, four hundred of the fourth, eleven hundred of the fifth, and three thousand two hundred of the sixth. We think we can easily number the stars; but train a six-inch telescope on a little section of the Twins, where six faint stars are visible, and over three thousand luminous points appear. The seventh magnitude has 13,000 stars; the eighth, 40,000; the ninth, 142,000. There are 18,000,000 stars in the zone called the Milky Way. When our eyes are not sensitive enough to be affected by the light of far-off stars the tasimeter feels their heat, and tells us the word of their Maker is true—"they are innumerable."*

Double and Multiple Stars.

If we look up during the summer months nearly overhead at the star ε Lyræ, east of Vega (Fig. 72), we shall see with the naked eye that the star appears a little

---

* *Telescopic Work.*—Look at the Hyades and Pleiades in Taurus. Notice the different colors of stars in them both. Find the cluster Præsepe in Fig. 70, just a trifle above a point midway between Procyon and Regulus. It is equally distant from Procyon and a point a little below Pollux. Sweep along the Milky Way almost anywhere, and observe the distribution of stars; in some places perfect crowds, in others more sparsely scattered. Find with the naked eye the rich cluster in Perseus. Draw a line from Algol to *a* of Perseus (Fig. 67); turn at right angles to the right, at a distance of once and four-tenths the first line a brightness will be seen. The telescope reveals a gorgeous cluster.

elongated. Turn your opera-glass upon it, and two stars appear. Turn a larger telescope on this double star, and each of the components separate into two. It is a double double star. We know that if two stars are near in reality, and not simply apparently so by being in the same line of sight, they must revolve around a common centre of gravity, or rush to a common ruin. Eagerly we watch to see if they revolve. A few years suffice to show them in actual revolution. Nay, the movement of revolution has been decided before the companion star was discovered. Sirius has long been known to have a proper motion, such as it would have if another sun were revolving about it. Even the direction of the unseen body could always be indicated. In February, 1862, Alvan Clark, artist, poet, and maker of telescopes (which requires even greater genius than to be both poet and artist), discovered the companion of Sirius just in its predicted place. As a matter of fact, one of Mr. Clark's sons saw it first; but their fame is one. The time of revolution of this pair is fifty years. But one companion does not meet the conditions of the movements. Here must also be one or more planets too small or dark to be seen. The double star ξ in the Great Bear (see Fig. 70) makes a revolution in fifty-eight years.

Procyon moves in an orbit which requires the presence of a companion star, but it has as yet eluded our search. Castor is a double star; but a third star or planet, as yet undiscovered, is required to account for its perturbations. Men who discovered Neptune by the perturbations of Uranus are capable of judging the cause of the perturbations of suns. We have spoken of the whole orbit of the earth being invisible from the stars. The nearest star in our northern hemisphere, 61 Cygni, is a telescopic double star; the constituent parts of it are forty-five times as far from each other as the earth is from the sun, yet it takes a large opera-glass to show any distance between the stars.*

When γ Virginis was observed in 1718 by Bradley, the component parts were 7″ asunder. He incidentally remarked in his note-book that the line of their connection was parallel to the line of the two stars Spica, or a and δ Virginis. By 1840 they were not more than 1″ apart, and the line of their connection greatly changed. The appearance of the star is given in Fig. 75 (15), commencing at the left, for the years 1837, '38, '39, '40, '45, '50, '60, and '79; also a conjectural orbit, placed obliquely, and the position of

---

* *Telescopic Work.*—Only such work will be laid out here as can be done by small telescopes of from two to four inch object-glasses. The numbers in Fig. 75 correspond to those of the table.

No.	Name.	Fig.	Dist. of Parts.	Magnitudes.	Remarks.
1.	ε Lyræ......	72	1′ 56″	......	Quadruple.
2.	ζ Lyræ.....	72	44	5 and 6	Topaz and green.
3.	β Cygni....	73	34½	3 " 6	Yellow and blue.
4.	61 Cygni...	73	20	5 " 6	Nearest star but one.
5.	Mizar.......	67	14	3 " 4	Both white.
6.	Polaris	67	18½	2 " 9	{Test object of eye and glass.
7.	ρ Orionis...	Frontispiece.	7	5 " 8	Yellow and blue.
8.	β Orionis...	"	9	1 " 8	Rigel.
9.	δ " ...	"	10	2 " 8	Red and white.
10.	θ " ...	"	......	......	Septuple.
11.	λ " ...	"	5		White and violet.
12.	σ " ...	" A, B.	11	4 " 10	Octuple.
13.	Castor......	69	5½	2 " 3	White.
14.	Pollux......	69	......	Triple.	Orange, gray, lilac.
15.	γ Virginis..	70	5	3 and 3	Both yellow.

the stars at the times mentioned, commencing at the top. The time of its complete revolution is one hundred and fifty years.

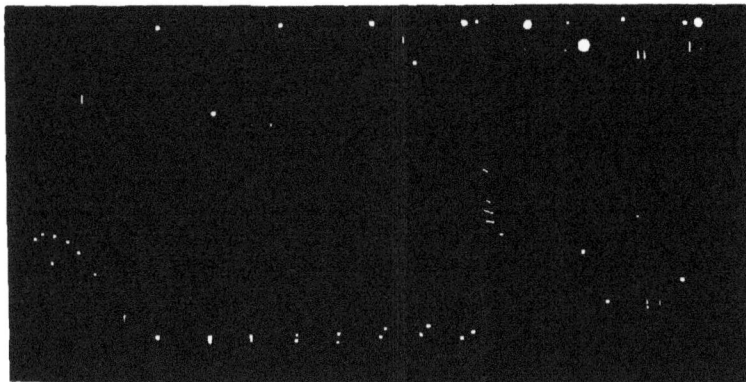

Fig. 75.—Aspects and Revolution of Double Stars.

The meaning of these double stars is that two or more suns revolve about their centre of gravity, as the moon and earth about their centre. If they have planets, as doubtless they have, the movement is no more complicated than the planets we call satellites of Saturn revolving about their central body, and also about the sun. Kindle Saturn and Jupiter to a blaze, or let out their possible light, and our system would appear a triple star in the distance. Doubtless, in the far past, before these giant planets were cooled, it so appeared.

We find some stars double, others triple, quadruple, octuple, and multiple. It is an extension of the same principles that govern our system. Some of these suns are so far asunder that they can swing their Neptunes between them, with less perturbation than Uranus and Neptune have in ours. Light all our planets, and there would be a multiple star with more or less suns seen, according to the power of the instrument. Perhaps the octuple star σ in Orion differs in no respect from our system, except in the size and distance of its separate bodies, and less cooling, either from being younger, or from the larger bodies cooling more slowly. Suns are of all ages. Infinite variety fills the sky. It is as preposterous to expect that every system or world should have analogous circumstances to ours at the present time, as to insist that every member of a family should be of the same age, and in the same state of development. There are worlds that have not yet reached the conditions of habitability by men, and worlds that have passed these conditions long since. Let them go. There are enough left, and an infinite number in the course of preparation. Some are fine and lasting enough to be eternal mansions.

Colored Stars.

In the smoky morning we get only red light, but the sun is white. So Aldebaran and Betelguese may be girt by vapors, that only the strong red rays can pass. Again, an iron moderately heated gives out dull red light; becoming hotter, it emits white light. Sirius, Regulus, Vega, and Spica may be white from greater intensity of vibration. Procyon, Capella, and Polaris are yellow from less intensity of vibration. Again, burn salt in a white flame, and it turns to yellow; mix alcohol and boracic acid, ignite them, and a beautiful green flame results; alcohol and nitrate of strontia give red flame; alcohol and nitrate of barytes give yellow flame. So the composition of a sun, or the special development of any one substance thereof at any time, may determine the color of a star.

The special glory of color in thé stars is seen in the marked contrasts presented in the double and multiple stars. The larger star is usually white, still in the intensity of heat and vibration; the others, smaller, are somewhat cooled off, and hence present colors lower down the scale of vibration, as green, yellow, orange, and even red.

That stars should change color is most natural. Many causes would produce this effect. The ancients said Sirius was red. It is now white. The change that would most naturally follow mere age and cooling would be from white, through various colors, to red. We are charmed with the variegated flowers of our gardens of earth, but he who makes the fields blush with flowers under the warm kisses of the sun has planted his wider gardens of space with colored stars. "The rainbow flowers of the footstool, and the starry flowers of the throne," proclaim one being as the author of them all.

Clusters of Stars.

From double and multiple we naturally come to groups and clusters. Allusion has been made to the Hyades, Pleiades, etc. Every one has noticed the Milky Way. It seems like two irregular streams of compacted stars. It is not supposed that they are necessarily nearer together than the stars in the sparse regions about the pole. But the 18,000,000 suns belonging to our system are arranged within a space represented by a flattened disk. If one hundred lights, three inches apart, are arranged on a hoop ten feet in diameter, they would be in a circle. Add a thousand or two more the same distance apart, filling up the centre, and extending a few inches on each side of the inner plane of the hoop: an eye in the centre, looking out toward the edge, would see a milky way of lights; looking out toward the sides or poles, would see comparatively few. It would seem as if this oblate spheroidal arrangement was the result of a revolution of all the suns composing the system. Jupiter and earth are flattened at the poles for the same reason.

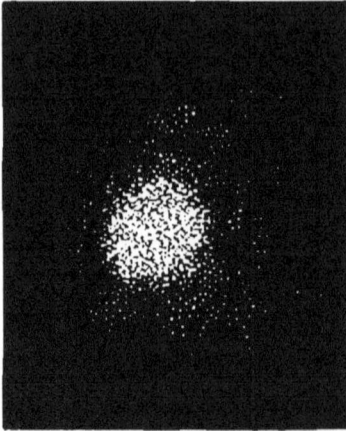

Fig. 76.—Sprayed Cluster below η in Hercules.

In various parts of the heavens there are small globular well-defined clusters, and clusters very irregular in form, marked with sprays of stars. There is a cluster of this latter class in Hercules, just under the S, in Fig. 72. "Probably no one ever saw it with a good telescope without a shout of wonder." Here is a cluster of the former class represented in Fig. 77. "The noble globular cluster ω Centauri is beyond all comparison the richest and largest object of the kind

Fig. 77.—Globular Cluster.

in the heavens. Its stars are literally innumerable; and as their total light, when received by the naked eye, affects it hardly more than a star of the fifth to fourth magnitude, the minuteness of each star may be imagined."

There are two possibilities of thought concerning these clusters. Either that they belong to our stellar system, and hence the stars must be small and young, or they are another universe of millions of suns, so far away that the inconceivable distances between the stars are shrunken to a hand's-breadth, and their unbearable splendor of innumerable suns can only make a gray haze at the distance at which we behold them. The latter is the older and grander thought; the former the newer and better substantiated.

Nebulæ.

The gorgeous clusters we have been considering appear to the eye or the small telescope as little cloudlets of hazy light. One after another were resolved into stars; and the natural conclusion was, that all would yield and reveal themselves to be clustered suns, when we had telescopes of sufficient power. But the spectroscope, seeing not merely form but substance also, shows that some of them are not stars in any sense, but masses of glowing gas. Two of these nebulæ are visible to the naked eye: one in Andromeda (see Fig. 68), and one around the middle star of the sword of Orion, shown in Fig. 78. A three-inch telescope resolves θ Orionis into the famous trapezium, and a five-inch instrument sees two stars more. The shape of the nebula is changeable, and is hardly suggestive of the moulding influence of gravitation. It is probably composed of glowing nitrogen and hydrogen gases. Nebulæ are of all conceivable shapes—circular, annular, oval, lenticular, conical, spiral, snake - like, looped, and nameless. Compare the sprays of the Crab nebulæ above ζ Tauri,

Fig. 78.—The great Nebula about the multiple Star θ Orionis. (See Frontispiece.)

seen in Fig. 79, and the ring nebula, Fig. 80. This last possibly consists of stars, and is situated, as shown in Fig. 81, midway between β and γ Lyræ.

When Herschel was sweeping the heavens with his telescope, and saw but few stars, he often said to his assistant, "Prepare to write; the nebulæ are coming." They are most abundant where the stars are least so. A zone about the heavens 30° wide, with the Milky Way in the centre, would include one-fourth of the celestial sphere; but instead of one-fourth, we find nine-tenths of the stars in this zone, and but one-tenth of the nebulæ.

These immense masses of unorganized matter are noticed to change their forms, vary their light greatly, but not quickly; they change through the ages. "God works slowly." He takes a thousand years to lift his hand off.

Fig. 79.—Crab Nebula, near ζ Tauri. (See Frontispiece.)

There are many unsolved problems connected with these strange bodies. Whether they belong to our system, or are beyond it, is not settled; the weight of evidence leans to the first view.

Variable Stars.

Our sun gives a variable amount of light, changing through a period of eleven years. Probably every star,

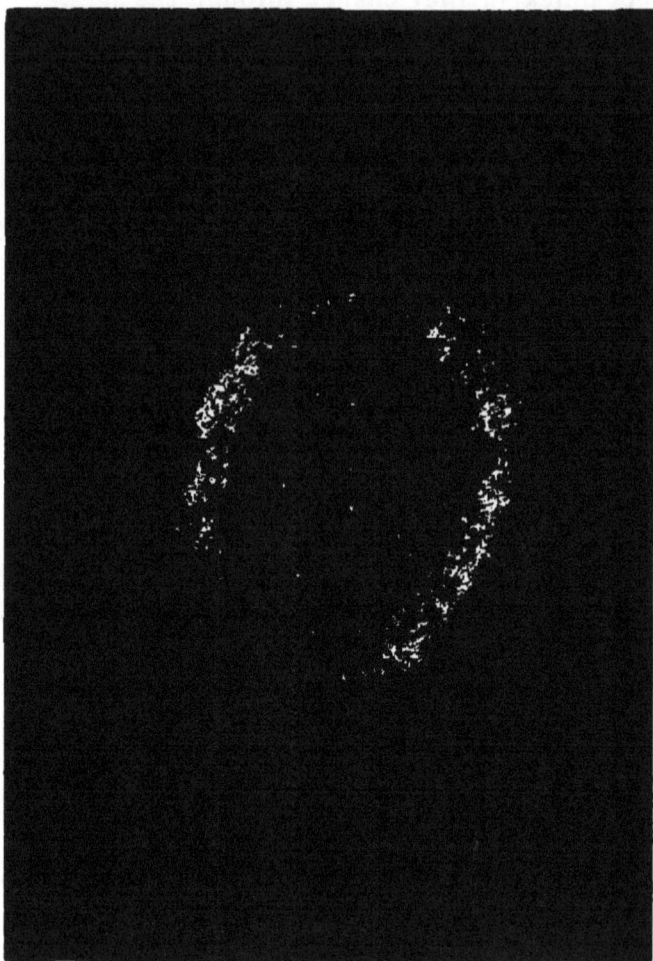

Fig. 80.—The Ring Nebula.

if examined by methods sufficiently delicate and exact, would be found to be variable. The variations of some

stars are so marked as to challenge investigation. β
Lyræ (Fig. 81) has two maxima and minima of light. In

Fig. 81.—Constellation Lyra, showing place of the Ring Nebula.

three days it rises from magnitude $4\frac{1}{2}$ to $3\frac{1}{2}$; in a week
it falls to 4, and rises to $3\frac{1}{2}$; and in three days more
drops to $4\frac{1}{2}$: it makes all these changes in thirteen
days; but this period is constantly increasing. The
variations of one hundred and forty-three stars have
been well ascertained.

Mira, or the Wonderful, in the Whale (Fig. 68), is
easily found when visible. Align from Capella to the
Pleiades, and as much farther, and four stars will be
seen, situated thus: • • •   The right-hand one is
Mira. For half a month it shines as a star of the second
magnitude. Then for three months it fades away, and
is lost to sight; going down even to the eleventh magnitude. But after five months its resurrection morning
comes; and in three months more—eleven months in
all—our Wonderful is in its full glory in the heavens.
But its period and brilliancy are also variable. The star
Megrez, δ in the Great Bear, has been growing dim

for a century.   In 1836 Betelguese was exceedingly
variable, and continued so till 1840, when the changes
became much less conspicuous.   Algol (Fig. 68) has
been already referred to.   This slowly winking eye
is of the second magnitude during 2d. 14h.   Then it
dozes off toward sleep for 4h. 24m., when it is nearly
invisible.   It wakes up during the same time; so that
its period from maximum brilliancy to the same state
again is 2d. 20h. 48m.   Its recognizable changes are
within five or six hours.   As I write, March 25th, 1879,
Algol gives its minimum light at 9h. 36m. P.M.   It
passes fifteen minima in 43d. 13m.   There will therefore
be another minimum May 7th, at 9h. 49m.   Its future
periods are easy to estimate.   Perhaps it has some dark
body revolving about it at frightful speed, in a period
of less than three days.   The period of its variability is
growing shorter at an increasing rate.   If its variability
is caused by a dark body revolving about it, the orbit of
that body is contracting, and the huge satellite will soon,
as celestial periods are reckoned, commence to graze the
surface of the sun itself, rebound again and again, and
at length plunge itself into the central fire.   Such an
event would evolve heat enough to make Algol flame
up into a star of the first magnitude, and perhaps out-
blaze Sirius or Capella in our winter sky.

None of the causes for these changes we have been
able to conjecture seem very satisfactory.   The stars
may have opaque planets revolving about them, shut-
ting off their light; they may rotate, and have unequal-
ly illuminated sides; they may revolve in very elliptical
orbits, so as to greatly alter their distance from us; they
may be so situated in regard to zones of meteorites as

to call down periodically vast showers; but none or all of these suppositions apply to all cases, if they do to any.

Temporary, New, and Lost Stars.

Besides regular movements to right and left, up and down, to and from us—changes in the intensity of illumination by changes of distance—besides variations occurring at regular and ascertainable intervals, there are stars called *temporary*, shining awhile and then disappearing; *new*, coming to a definite brightness, and so remaining; and *lost*, those whose first appearance was not observed, but which have utterly disappeared.

In November, 1572, a new star blazed out in Cassiopeia. Its place is shown in Fig. 67, χ γ being the stars in the seat of the chair, and δ being the first one in the back. This star was visible at noonday, and was brighter than any other star in the heavens. In January, 1573, it was less bright than Jupiter; in April it was below the second magnitude, and the last of May it utterly disappeared. It was as variable in color as in brilliancy. During its first two months, the period of greatest brightness, it was dazzling white, then became yellow, and finally as red as Mars or Aldebaran, and so expired.

A bright star was seen very near to the place of the *Pilgrim*, as the star of 1572 was called, in A.D. 945 and 1264. A star of the tenth magnitude is now seen brightening slowly almost exactly in the same place. It is possible that this is a variable star of a period of about three hundred and ten years, and will blaze out again about 1885.

But we have had, within a few years, fine opportunities to study, with improved instruments, two new stars. On the evening of May 12th, 1866, a star of the second magnitude was observed in the Northern Crown, where no star above the fifth magnitude had been twenty-four hours before. In Argelander's chart a star of the tenth magnitude occupies the place. May 13th it had declined to the third magnitude, May 16th to the fourth, May 17th to the fifth, May 19th to the seventh, May 31st to the ninth, and has since diminished to the tenth. The spectroscope showed it to be a star in the usual condition; but through the usual colored spectrum, crossed with bright lines, shone four bright lines, two of which indicated glowing hydrogen. Here was plenty of proof that an unusual amount of this gas had given this sun its sudden flame. As the hydrogen burned out the star grew dim.

Two theories immediately presented themselves: First, that vast volumes had been liberated from within the orb by some sudden breaking up of the doors of its great deeps; or, second, this star had precipitated upon itself, by attraction, some other sun or planet, the force of whose impact had been changed into heat.

Though we see the liberated hydrogen of our sun burst up with sudden flame, it can hardly be supposed that enough could be liberated at once to increase the light and heat one hundred-fold.

In regard to the second theory, it is capable of proof that two suns half as large as ours, moving at a velocity of four hundred and seventy-six miles per second, would evolve heat enough to supply the radiation of our sun for fifty million years. How could it be possible for a sun like this newly blazing orb to cool off to such a degree in a month? Besides, there would not be one chance in a thousand for two orbs to come directly together. They would revolve about each other till a kind of grazing contact of grinding worlds would slowly kindle the ultimate heat.

It is far more likely that this star encountered an enormous stream of meteoric bodies, or perhaps absorbed a whole comet, that laid its million leagues of tail as fuel on the central fire. Only let it be remembered that the fuel is far more force than substance. Allusion has already been made to the sudden brightening of our sun on the first day of September, 1859. That was caused, no doubt, by the fall of large meteors, following in the train of the comet of 1843, or some other comet. What the effect would have been, had the whole mass of the comet been absorbed, cannot be imagined.

Another new star lately appeared in Cygnus, near the famous star 61—the first star in the northern hemisphere whose distance was determined. It was first seen November 24th, 1876, as a third magnitude star of a yellow color. By December 2d it had sunk to the fourth magnitude, and changed to a greenish color. It had then three bright hydrogen lines, the strong double sodium line, and others, which made it strongly resemble the spectrum of the chromosphere of our sun. An entirely different result appeared in the fading of these two stars. In the case of the star in the Crown, the extraordinary light was the first to fade, leaving the usual stellar spectrum. In the case of the star in Cygnus, the part of the spectrum belonging to stellar light was the first to fade, leaving the bright lines; that is, the gas of one gave way to regular starlight, and the starlight of the other having faded, the regular light of the glowing gas continued. By some strange oversight, no one studied the star again for six months. In September and November, 1877, the light of this star was found to be blue, and not to be starlight at all. It had no rainbow spectrum, only one kind of rays, and hence only one color. Its sole spectroscopic line is believed to be that of glowing nitrogen gas. We have then, probably, in the star of 1876, a body shining by a feeble and undiscernible light, surrounded by a discernible immensity of light of nitrogen gas. This is its usual condition; but if a flight of meteors should raise the heat of the central body so as to outshine the nebulous envelope, we should have the conditions we discovered in November, 1876. But a rapid cooling dissipates the observable light of all colors, and leaves only the glowing gas of one color.

Movements of Stars.

We call the stars *fixed*, but motion and life are necessary to all things. Besides the motion in the line of sight described already, there is motion in every other conceivable direction. We knew Sirius moved before we had found the cause. We know that our sun moves back and forth in his easy bed one-half his vast diameter, as the larger planets combine their influence on one side or the other.

The sun has another movement. We find the stars in Hercules gradually spreading from each other. Hercules's brawny limbs grow brawnier every century. There can be but one cause: we are approaching that quarter of the heavens. (See ⊕, Fig. 72.) We are even able to compute the velocity of our approach; it is eight miles a second. The stars in the opposite quarter of the heavens in the Dove are drawing nearer together.

This movement would have no effect on the apparent place of the stars at either pole, if they were all equally distant; but it must greatly extend or contract the apparent space between them, since they are situated at various distances.

Independent of this, the stars themselves are all in motion, but so vast is the distance from which we observe them that it has taken an accumulation of centuries before they could be made measurable. A train going forty miles an hour, seen from a distance of two miles, almost seems to stand still. Arcturus moves through space three times as fast as the earth, but it takes a century to appear to move the eighth part of the diameter of the moon. There is a star in the Hunting Dogs, known as 1830 Groombridge, which has a velocity beyond what all the attraction of the matter of the known universe could give it. By the year 9000 it may be in Berenice's Hair.

Some stars have a common movement, being evidently related together. A large proportion of the brighter stars between Aldebaran and the Pleiades have a common motion eastward of about ten seconds a century. All the angles marked by a, β, γ, χ Orionis will be altered in different directions; λ is moving toward γ. Λ and ε will appear as a double star. In A.D. 50,000 Procyon will be nearer χ Orionis than Rigel now is, and Sirius will be in line with a and χ Orionis. All the stars of the Great Dipper, except Benetnasch and Dubhe, have a common motion somewhat in the direction of Thuban (Fig. 67), while the two named have a motion nearly opposite. In 36,000 years the end of the Dipper will have fallen out so that it will hold no water, and the handle will be broken square off at Mizar. "The Southern Cross," says Humboldt, "will not always keep its characteristic form, for its four stars travel in different directions with unequal velocities. At the present time it is not known how many myriads of years must elapse before its entire dislocation."

These movements are not in fortuitous or chaotic ways, but are doubtless in accordance with some perfect plan. We have climbed up from revolving earth and moon to revolving planets and sun, in order to understand how two or ten suns can revolve about a common centre. Let us now leap to the grander idea that all the innumerable stars of a winter night not only can, but must revolve about some centre of gravity. Men have been looking for a central sun of suns, and have not found it. None is needed. Two suns can balance about a point; all suns can swing about a common centre. That one unmoving centre may be that city more gorgeous than Eastern imagination ever conceived, whose pavement is transparent gold, whose walls are precious stones, whose light is life, and where no dark planetary bodies ever cast shadows. There reigns the King and Lord of all, and ranged about are the far-off provinces of his material systems. They all move in his sight, and receive power from a mind that never wearies.

XI.

THE WORLDS AND THE WORD.

"The worlds were framed by the word of God."—*Heb*. xi., 3.

" Mysterious night! when our first parent knew thee
From report divine, and heard thy name,
Did he not tremble for this lovely frame,
This glorious canopy of light and blue?
Yet, 'neath a curtain of translucent dew,
Bathed in the rays of the great setting flame,
Hesperus, with all the host of heaven, came,
And lo! creation widened in man's view.
Who could have thought such darkness lay concealed
Within thy beams, O Sun! Oh who could find,
Whilst fruit and leaf and insect stood revealed,
That to such countless worlds thou mad'st us blind!
Why do we then shun death with anxious strife?
If light conceal so much, wherefore not life?"

BLANCO WHITE

XI.

THE WORLDS AND THE WORD.

MEN have found the various worlds to be far richer than they originally thought. They have opened
door after door in their vast treasuries, have ascended
throne after throne of power, and ruled realms of increasing extent. We have no doubt that unfoldings
in the future will amaze even those whose expectations
have been quickened by the revealings of the past.
What if it be found that the Word is equally inexhaustible?.

After ages of thought and discovery we have come
out of the darkness and misconceptions of men. We
believe in no serpent, turtle, or elephant supporting
the world; no Atlas holding up the heavens; no crystal domes, " with cycles and epicycles scribbled o'er."
What if it be found that one book, written by ignorant
men, never fell into these mistakes of the wisest! Nay,
more, what if some of the greatest triumphs of modern
science are to be found plainly stated in a book older
than the writings of Homer? If suns, planets, and
satellites, with all their possibilities of life, changes of
flora and fauna, could be all provided for, as some
scientists tell us, in the fiery star-dust of a cloud, why
may not the same Author provide a perpetually widening river of life in his Word? As we believe He is
perpetually present in his worlds, we know He has

promised to be perpetually present in his Word, making it alive with spirit and life.

The wise men of the past could not avoid alluding to ideas the falsity of which subsequent discovery has revealed; but the writers of the Bible did avoid such erroneous allusion. Of course they referred to some things, as sunrise and sunset, according to appearance; but our most scientific books do the same to-day. That the Bible could avoid teaching the opposite of scientific truth proclaims that a higher than human wisdom was in its teaching.

That negative argument is strong, but the affirmative argument is much stronger. The Bible declares scientific truth far in advance of its discovery, far in advance of man's ability to understand its plain declarations. Take a few conspicuous illustrations:

The Bible asserted from the first that the present order of things had a beginning. After ages of investigation, after researches in the realms of physics, arguments in metaphysics, and conclusions by the necessities of resistless logic, science has reached the same result.

The Bible asserted from the first that creation of matter preceded arrangement. It was chaos—void—without form—darkness; arrangement was a subsequent work. The world was not created in the form it was to have; it was to be moulded, shaped, stratified, coaled, mountained, valleyed, subsequently. All of which science utters ages afterward.

The Bible did not hesitate to affirm that light existed before the sun, though men did not believe it, and used it as a weapon against inspiration. Now we praise men for having demonstrated the oldest record.

It is a recently discovered truth of science that the strata of the earth were formed by the action of water, and the mountains were once under the ocean. It is an idea long familiar to Bible readers: "Thou coverest the earth with the deep as with a garment. The waters stood above the mountains. At thy rebuke they fled; at the voice of thy thunder they hasted away. The mountains ascend; the valleys descend into the place thou hast founded for them." Here is a whole volume of geology in a paragraph. The thunder of continental convulsions is God's voice; the mountains rise by God's power; the waters haste away unto the place God prepared for them. Our slowness of geological discovery is perfectly accounted for by Peter. "For of this they are *willingly ignorant*, that by the word of God there were heavens of old, and land framed out of water, and by means of water, whereby the world that then was, being overflowed by water, perished." We recognize these geological subsidences, but we read them from the testimony of the rocks more willingly than from the testimony of the Word.

Science exults in having discovered what it is pleased to call an order of development on earth—tender grass, herb, tree; moving creatures that have life in the waters; bird, reptile, beast, cattle, man. The Bible gives the same order ages before, and calls it God's successive creations.

During ages on ages man's wisdom held the earth to be flat. Meanwhile, God was saying, century after century, of himself, "He sitteth upon the sphere of the earth" (Gesenius).

Men racked their feeble wits for expedients to up-

hold the earth, and the best they could devise were serpents, elephants, and turtles; beyond that no one had
ever gone to see what supported them. Meanwhile,
God was perpetually telling men that he had hung the
earth upon nothing.

Men were ever trying to number the stars. Hippar-
chus counted one thousand and twenty-two; Ptolemy
one thousand and twenty-six; and it is easy to number
those visible to the naked eye. But the Bible said,
when there were no telescopes to make it known, that
they were as the sands of the sea, "innumerable." Science has appliances of enumeration unknown to other
ages, but the space-penetrating telescopes and tastime-
ters reveal more worlds—eighteen millions in a single
system, and systems beyond count—till men acknowl-
edge that the stars are innumerable to man. It is
God's prerogative "to number all the stars; he also
calleth them all by their names."

Torricelli's discovery that the air had weight was received with incredulity. For ages the air had propelled
ships, thrust itself against the bodies of men, and over-
turned their works. But no man ever dreamed that
weight was necessary to give momentum. During all
the centuries it had stood in the Bible, waiting for
man's comprehension: "He gave to the air its weight"
(Job xxviii. 25).

The pet science of to-day is meteorology. The fluc-
tuations and variations of the weather have hitherto
baffled all attempts at unravelling them. It has seemed
that there was no law in their fickle changes. But at
length perseverance and skill have triumphed, and a
single man in one place predicts the weather and winds

for a continent. But the Bible has always insisted that the whole department was under law; nay, it laid down that law so clearly, that if men had been willing to learn from it they might have reached this wisdom ages ago. The whole moral law is not more clearly crystallized in "Thou shalt love the Lord thy God with all thy heart, and thy neighbor as thyself," than all the fundamentals of the science of meteorology are crystallized in these words: "The wind goeth toward the south (equator), and turneth about (up) unto the north; it whirleth about continually, and the wind returneth again according to his circuits (established routes). All the rivers run into the sea; yet the sea is not full: unto the place from whence the rivers come, thither they return again" (Eccles. i. 6, 7).

. Those scientific queries which God propounded to Job were unanswerable then; most of them are so now. "Whereon are the sockets of the earth made to sink?" Job never knew the earth turned in sockets; much less could he tell where they were fixed. God answered this question elsewhere. "He stretcheth the north (one socket) over the empty place, and hangeth the earth upon nothing." Speaking of the day-spring, God says the earth is *turned* to it, as clay to the seal. The earth's axial revolution is clearly recognized. Copernicus declared it early; God earlier.

No man yet understands the balancing of the clouds, nor the suspension of the frozen masses of hail, any more than Job did.

Had God asked if he had perceived the *length* of the earth, many a man to-day could have answered yes. But the eternal ice keeps us from perceiving the *breadth* of the earth, and shows the discriminating wisdom of the question.

The statement that the sun's going is from the end of the heaven, and his circuit to the ends of it, has given edge to many a sneer at its supposed assertion that the sun went round the earth. It teaches a higher truth— that the sun itself obeys the law it enforces on the planets, and flies in an orbit of its own, from one end of heaven in Argo to the other in Hercules.

So eminent an astronomer and so true a Christian as General Mitchell, who understood the voices in which the heavens declare the glory of God, who read with delight the Word of God embodied in worlds, and who fed upon the written Word of God as his daily bread, declared, " We find an aptness and propriety in all these astronomical illustrations, which are not weakened, but amazingly strengthened, when viewed in the clear light of our present knowledge." Herschel says, "All human discoveries seem to be made only for the purpose of confirming more strongly the truths that come from on high, and are contained in the sacred writings." The common authorship of the worlds and the Word becomes apparent; their common unexplorable wealth is a necessary conclusion.

Since the opening revelations of the past show an unsearchable wisdom in the Word, has that Word any prophecy concerning mysteries not yet understood, and events yet in the future? There are certain problems as yet insolvable. We have grasped many clews, and followed them far into labyrinths of darkness, but not yet through into light.

We ask in vain, "What is matter?" No man can

answer. We trace it up through the worlds, till its increasing fineness, its growing power, and possible identity of substance, seem as if the next step would reveal its spirit origin. What we but hesitatingly stammer, the Word boldly asserts.

We ask, "What is force?" No man can answer. We recognize its various grades, each subordinate to the higher—cohesion dissolvable by heat; the affinity of oxygen and hydrogen in water overcome by the piercing intensity of electric fire; rivers seeking the sea by gravitation carried back by the sun; rock turned to soil, soil to flowers; and all the forces in nature measurably subservient to mind. Hence we partly understand what the Word has always taught us, that all lower forces must be subject to that which is highest. How easily can seas be divided, iron made to swim, water to burn, and a dead body to live again, if the highest force exert itself over forces made to be mastered. When we have followed force to its highest place, we always find ourselves considering the forces of mind and spirit, and say, in the words of the Scriptures, "God is spirit."

We ask in vain what is the end of the present condition of things. We have read the history of our globe with great difficulty—its prophecy is still more difficult. We have asked whether the stars form a system, and if so, whether that system is permanent. We are not able to answer yet. We have said that the sun would in time become as icy cold and dead as the moon, and then the earth would wander darkling in the voids of space. But the end of the earth, as prophesied in the Word, is different: "The heavens will pass away with a rushing noise, and the elements will be dissolved with burning heat, and the earth and the works therein will be burned up." The latest conclusions of science point the same way. The great zones of uncondensed matter about the sun seem to constitute a resisting medium as far as they reach. Encke's comet, whose orbit comes near the sun, is delayed. This gives gravitation an overwhelming power, and hence the orbit is lessened and a revolution accomplished more quickly. Faye's comet, which wheels beyond the track of Mars, is not retarded. If the earth moves through a resisting substance, its ultimate fall into the sun is certain. Whether in that far future the sun shall have cooled off, or will be still as hot as to-day, Peter's description would admirably portray the result of the impact. Peter's description, however, seems rather to indicate an interference of Divine power at an appropriate time before a running down of the system at present in existence, and a re-endowment of matter with new capabilities.

After thousands of years, science discovered the true way to knowledge. It is the Baconian way of experiment, of trial, of examining the actual, instead of imagining the ideal. It is the acceptance of the Scriptural plan. "If a man wills to do God's will, he shall know." Oh taste and see! In science men try hypotheses, think the best they can, plan broadly as possible, and then see if facts sustain the theory. They have adopted the Scriptural idea of accepting a plan, and then working in faith, in order to acquire knowledge. Fortunately, in the work of salvation the plan is always perfect. But, in order to make the trial under the most favorable circumstances, there must be faith. The faith of

science is amazing; its assertions of the supersensual
are astounding. It affirms a thousand things that cannot be physically demonstrated: that the flight of a rifle-
ball is parabolic; that the earth has poles; that gases are
made of particles; that there are atoms; that an electric light gives ten times as many rays as are visible;
that there are sounds to which we are deaf, sights to
which we are blind; that a thousand objects and activities are about us, for the perception of which we need a
hundred senses instead of five. These faiths have nearly all led to sight; they have been rewarded, and the
world's wealth of knowledge is the result. The Word
has ever asserted the supersensuous, solicited man's faith,
and ever uplifted every true faith into sight. Lowell is
partly right when he sings:

> "Science was Faith once; Faith were science now,
> Would she but lay her bow and arrows by,
> And arm her with the weapons of the time."

Faith laid her bow and arrows by before men in pursuit
of worldly knowledge discovered theirs.

What becomes of the force of the sun that is being
spent to-day? It is one of the firmest rocks of science
that there can be no absolute destruction of force. It
is all conserved somehow. But how? The sun con-
tracts, light results, and leaps swiftly into all encircling
space. It can never be returned. Heat from stars invisible by the largest telescope enters the tastimeter, and
declares that that force has journeyed from its source
through incalculable years. There is no encircling
dome to reflect all this force back upon its sources. Is
it lost? Science, in defence of its own dogma, should

assign light a work as it flies in the space which we have learned cannot be empty. There ought to be a realm where light's inconceivable energy is utilized in building a grander universe, where there is no night. Christ said, as he went out of the seen into the unseen, "I go to prepare a place for you;" and when John saw it in vision the sun had disappeared, the moon was gone, but the light still continued.

Science finds matter to be capable of unknown refinement; water becomes steam full of amazing capabilities: we add more heat, superheat the steam, and it takes on new aptitudes and uncontrollable energy. Zinc burned in acid becomes electricity, which enters iron as a kind of soul, to fill all that body with life. All matter is capable of transformation, if not transfiguration, till it shines by the light of an indwelling spirit. Scripture readers know that bodies and even garments can be transfigured, be made ἀστράπτων (Luke xxiv. 4), shining with an inner light. They also look for new heavens and a new earth endowed with higher powers, fit for perfect beings.

When God made matter, so far as our thought permits us to know, he simply made force stationary and unconscious. Thereafter he moves through it with his own will. He can at any time change these forces, making air solid, water and rock gaseous, a world a cloud, or a fire-mist a stone. He may at some time restore all force to consciousness again, and make every part of the universe thrill with responsive joy. "Then shall the mountains and the hills break forth before you into singing, and all the trees of the field clap their hands." One of these changes is to come to the earth.

Amidst great noise the heaven shall flee, the earth be
burned up, and all their forces be changed to new
forms. Perhaps it will not then be visible to mortal
eyes. Perhaps force will then be made conscious, and
the flowers thereafter return our love as much as lower
creatures do now. A river and tree of life may be
consciously alive, as well as give life. Poets that are
nearest to God are constantly hearing the sweet voices
of responsive feeling in nature.

> "For his gayer hours
> She has a voice of gladness and a smile,
> And eloquence of beauty; and she glides
> Into his darker musings with a mild
> And gentle sympathy, that steals away
> Their sharpness ere he is aware."

Prophets who utter God's voice of truth say, "The
wilderness and the solitary place shall be glad for holy
men, and the desert shall rejoice and blossom as the
rose. It shall blossom abundantly and rejoice, even
with joy and singing."

Distinguish clearly between certainty and surmise.
The certainty is that the world will pass through catas-
trophic changes to a perfect world. The grave of uniformitarianism is already covered with grass. He that
creates promises to complete. The invisible, impon-
derable, inaudible ether is beyond our apprehension;
it transmits impressions 186,000 miles a second; it is
millions of times more capable and energetic than air.
What may be the bounds of its possibility none can imagine, for law is not abrogated nor designs disregarded
as we ascend into higher realms. Law works out more
beautiful designs with more absolute certainty. Why

11

should there not be a finer universe than this, and disconnected from this world altogether—a fit home for immortal souls? It is a necessity.

God filleth all in all, is everywhere omnipotent and wise. Why should there be great vacuities, barren of power and its creative outgoings? God has fixed the stars as proofs of his agency at some points in space. But is it in points only? Science is proud of its discovery that what men once thought to be empty space is more intensely active than the coarser forms of matter can be. But in the long times which are past Job glanced at earth, seas, clouds, pillars of heaven, stars, day, night, all visible things, and then added: "Lo! these are only the outlying borders of his works. What a whisper of a word we hear of *Him!* The thunder of his power who can comprehend?"

Science discovers that man is adapted for mastery in this world. He is of the highest order of visible creatures. Neither is it possible to imagine an order of beings generically higher to be connected with the conditions of the material world. This whole secret was known to the author of the oldest writing. "And God blessed them, and God said unto them: Be fruitful, and multiply, and replenish the earth, and subdue it: and have dominion over the fish of the sea, and over the fowl of the air, and over every living thing that moveth upon the earth." The idea is never lost sight of in the sacred writings. And while every man knows he must fail in one great contest, and yield himself to death, the later portions of the divine Word offer him victory even here. The typical man is commissioned to destroy even death, and make man a sharer in the victory.

Science babbles at this great truth of man's position like a little child; Scripture treats it with a breadth of perfect wisdom we are only beginning to grasp.

Science tells us that each type is prophetic of a higher one. The whale has bones prophetic of a human hand. Has man reached perfection? Is there no prophecy in him? Not in his body, perhaps; but how his whole soul yearns for greater beauty. As soon as he has found food, the savage begins to carve his paddle, and make himself gorgeous with feathers. How man yearns for strength, subduing animal and cosmic forces to his will! How he fights against darkness and death, and strives for perfection and holiness! These prophecies compel us to believe there is a world where powers like those of electricity and luminiferous ether are ever at hand; where its waters are rivers of life, and its trees full of perfect healing, and from which all unholiness is forever kept. What we infer, Scripture affirms.

Science tells us there has been a survival of the fittest. Doubtless this is so. So in the future there will be a survival of the fittest. What is it? Wisdom, gentleness, meekness, brotherly kindness, and charity. Over those who have these traits death hath no permanent power. The caterpillar has no fear as he weaves his own shroud; for there is life within fit to survive, and ere long it spreads its gorgeous wings, and flies in the air above where once it crawled. Man has had two states of being already. One confined, dark, peculiarly nourished, slightly conscious; then he was born into another—wide, differently nourished, and intensely conscious. He knows he may be born again into a life

wider yet, differently nourished, and even yet more intensely conscious.   Science has no hint how a long
ascending series of developments crowned by man may
advance another step, and make man ἰσάγγελος—equal
to angels.   But the simplest teaching of Scripture points
out a way so clear that a child need not miss the glorious consummation.

When Uranus hastened in one part of its orbit,
and then retarded, and swung too wide, men said there
must be another attracting world beyond; and, looking
there, Neptune was found.   So, when individual men
are so strong that nations or armies cannot break down
their wills; so brave, that lions have no terrors; so
holy, that temptation cannot lure nor sin defile them;
so grand in thought, that men cannot follow; so pure
in walk, that God walks with them—let us infer an attracting world, high and pure and strong as heaven.
The eleventh chapter of Hebrews is a roll-call of heroes
of whom this world was not worthy.   They were tort-
ured, not accepting deliverance, that they might obtain
a better resurrection.   The world to come influenced,
as it were, the orbits of their souls, and when their bodies fell off, earth having no hold on them, they sped on
to their celestial home.   The tendency of such souls
necessitates such a world.

The worlds and the Word speak but one language,
teach but one set of truths.   How was it possible that
the writers of the earlier Scriptures described physical
phenomena with wonderful sublimity, and with such
penetrative truth?   They gazed upon the same heaven
that those men saw who ages afterward led the world
in knowledge.   These latter were near-sighted, and ab-

sorbed in the pictures on the first veil of matter; the former were far-sighted, and penetrated a hundred strata of thickest material, and saw the immaterial power behind. The one class studied the present, and made the gravest mistakes; the other pierced the uncounted ages of the past, and uttered the profoundest wisdom. There is but one explanation. He that planned and made the worlds inspired the Word.

Science and religion are not two separate departments, they are not even two phases of the same truth. Science has a broader realm in the unseen than in the seen, in the source of power than in the outcomes of power, in the sublime laws of spirit than in the laws of matter; and religion sheds its beautiful light over all stages of life, till, whether we eat or whether we drink, or whatsoever we do, we may do all for the glory of God. Science and religion make common confession that the great object of life is to learn and to grow. Both will come to see the best possible means, for the attainment of this end is a personal relation to a teach er who is the Way, the Truth, and the Life.

XII.

THE ULTIMATE FORCE.

"In the beginning was the Word, and the Word was with God, and the Word was God. The same was in the beginning with God. All things became by him, and without him was not anything made that was made * * * and by him all things stand together."

"O thou eternal one; whose presence bright
All space doth occupy—all motion guide—
Thou from primeval nothingness didst call
First chaos, then existence.   Lord, on thee
Eternity had its foundation: all
Sprung forth from thee—of light, joy, harmony,
Sole origin: all life, all beauty thine.
Thy word created all, and doth create;
Thy splendor fills all space with rays divine;
Thou art and wert, and shalt be glorious, great.
Life-giving, life-sustaining Potentate,
Thy chains the unmeasured universe surround—
Upheld by thee, by thee inspired with breath."

DERZHAVIN

XII.

THE ULTIMATE FORCE.

THE universe is God's name writ large. Thought goes up the shining suns as golden stairs, and reads the consecutive syllables—all might, and wisdom, and beauty; and if the heart be fine enough and pure enough, it also reads everywhere the mystic name of love. Let us learn to read the hieroglyphics, and then turn to the blazonry of the infinite page. That is the key-note; the heavens and the earth declaring the glory of God, and men with souls attuned listening.

To what voices shall we listen first? Stand on the shore of a lake set like an azure gem among the bosses of green hills. The patter of rain means an annual fall of four cubic feet of water on every square foot of it. It weighs sixty-two and three-tenths pounds to the cubic foot, *i.e.*, fifty-two million tons on the surface of a little sheet of water twenty miles long by three wide. Now, all that weight of falling rain had to be lifted, a work compared to which taking up mountains and casting them into the sea is pastime. All that water had to be taken up before it could be cast down, and carried hundreds of miles before it could be there. You have heard Niagara's thunder; have stood beneath the falling immensity; seen it ceaselessly poured from an infinite hand; felt that you would be ground to atoms if you fell into that resistless flood. Well, all that infinity of water had to be lifted by main force, had to be taken up out of the far Pacific, brought over the Rocky Mountains; and the Mississippi keeps bearing its wide miles of water to the Gulf, and Niagara keeps thundering age after age, because there is power somewhere to carry the immeasurable floods all the time the other way in the upper air.

But this is only the Alpha of power. Professor Clark, of Amherst, Massachusetts, found that such a soft and pulpy thing as a squash had so great a power of growth that it lifted three thousand pounds, and held it day and night for months. It toiled and grew under the growing weight, compacting its substance like oak to do the work. All over the earth this tremendous power and push of life goes on—in the little star-eyed flowers that look up to God only on the Alpine heights, in every tuft of grass, in every acre of wheat, in every mile of prairie, and in every lofty tree that wrestles with the tempests of one hundred winters. But this is only the B in the alphabet of power.

Rise above the earth, and you find the worlds tossed like playthings, and hurled seventy times as fast as a rifle-ball, never an inch out of place or a second out of time. But this is only the C in the alphabet of power.

Rise to the sun. It is a quenchless reservoir of high-class energy. Our tornadoes move sixty miles an hour, those of the sun twenty thousand miles an hour. A forest on fire sends its spires of flame one hundred feet in air, the sun sends its spires of flame two hundred thousand miles. All our fires exhaust the fuel and burn out. If the sun were pure coal, it would burn out in five thousand years; and yet this sea of unquenchable

flame seethes and burns, and rolls and vivifies a dozen
worlds, and flashes life along the starry spaces for a million years without any apparent diminution. It sends
out its power to every planet, in the vast circle in which
it lies. It fills with light not merely a whole circle, but
a dome; not merely a dome above, but one below, and
on every side. At our distance of ninety-two and a half
millions of miles, the great earth feels that power in
gravitation, tides, rains, winds, and all possible life—
every part is full of power. Fill the earth's orbit with
a circle of such receptive worlds—seventy thousand instead of one—every one would be as fully supplied with
power from this central source. More. Fill the whole
dome, the entire extent of the surrounding sphere, bottom, sides, top, a sphere one hundred and eighty-five
million miles in diameter, and every one of these uncountable worlds would be touched with the same power as one; each would thrill with life. This is only the
D of the alphabet of power. And glancing up to the
other suns, one hundred, five hundred, twelve hundred
times as large, double, triple, septuple, multiple suns, we
shall find power enough to go through the whole alphabet in geometrical ratio; and then in the clustered suns,
galaxies, and nebulæ, power enough still unrepresented
by single letters to require all combinations of the alphabet of power. What is the significance of this single
element of power? The answer of science to-day is "correlation," the constant evolution of one force from another. Heat is a mode of motion, motion a result of heat.
So far so good. But are we mere reasoners in a circle?
Then we would be lost men, treading our round of death
in a limitless forest. What is the ultimate? Reason

out in a straight line. No definition of matter allows it
to originate force; only mind can do that. Hence the
ultimate force is always mind. Carry your correlation
as far as you please—through planets, suns, nebulæ, con-
cretionary vortices, and revolving fire-mist—there must
always be mind and will beyond. Some of that will-
power that works without exhaustion must take its own
force and render it static, apparent. It may do this in
such correlated relation that that force shall go on year
after year to a thousand changing forms; but that force
must originate in mind.

Go out in the falling rain, stand under the thunderous Niagara, feel the immeasurable rush of life, see the
hanging worlds, and trace all this—the carried rain, the
terrific thunder with God's bow of peace upon it, and
the unfalling planets hung upon nothing—trace all this
to the orb of day blazing in perpetual strength, but stop
not there. Who *made* the sun? Contrivance fills all
thought. *Who* made the sun? Nature says there is a
mind, and that mind is Almighty. Then you have read
the first syllables, viz., being and power.

What is the continuous relation of the universe to the
mind from which it derived its power? Some say that
it is the relation of a wound-up watch to the winder. It
was dowered with sufficient power to revolve its ceaseless changes, and its maker is henceforth an absentee
God. Is it? Let us have courage to see. For twenty years one devotes ten seconds every night to putting a little force into a watch. It is so arranged that
it distributes that force over twenty - four hours. In
that twenty years more power has been put into that
watch than a horse could exert at once. But suppose

one had tried to put all that force into the watch at once: it would have pulverized it to atoms. But supposing the universe had been dowered with power at first to run its enormous rounds for twenty millions of years. It is inconceivable; steel would be as friable as sand, and strengthless as smoke, in such strain.

We have discovered some of the laws of the force we call gravitation. But what do we know of its essence? How it appears to act we know a little, what it is we are profoundly ignorant. Few men ever discuss this question. All theories are sublimely ridiculous, and fail to pass the most primary tests. How matter can act where it is not, and on that with which it has no connection, is inconceivable.

Newton said that any one who has in philosophical matters a competent faculty of thinking, could not admit for a moment the possibility of a sun reaching through millions of miles, and exercising there an attractive power. A watch may run if wound up, but how the watch-spring in one pocket can run the watch in another is hard to see. A watch is a contrivance for distributing a force outside of itself, and if the universe runs at all on that principle, it distributes some force outside of itself.

Le Sage's theory of gravitation by the infinitive hail of atoms cannot stand a minute, hence we come back as a necessity of thought to Herschel's statement. "It is but reasonable to regard gravity as a result of a consciousness and a will existent somewhere." Where? I read an old book speaking of these matters, and it says of God, He hangeth the earth upon nothing; he upholdeth constantly all things by the word of his power

By him all things consist or hold together. It teaches an imminent mind; an almighty, constantly exerted power. Proof of this starts up on every side. There is a recognized tendency in all high-class energy to deteriorate to a lower class. There is steam in the boiler, but it wastes without fuel. There is electricity in the jar, but every particle of air steals away a little, unless our conscious force is exerted to regather it. There is light in the sun, but infinite space waits to receive it, and takes it swift as light can leap. We said that if the sun were pure coal, it would burn out in five thousand years, but it blazes undimmed by the million. How can it? There have been various theories: chemical combustion, it has failed; meteoric impact, it is insufficient; condensation, it is not proved; and if it were, it is an intermediate step back to the original cause of condensation. The far-seeing eyes see in the sun the present active power of Him who first said, "Let there be light," and who at any moment can meet a Saul in the way to Damascus with a light above the brightness of the sun—another noon arisen on mid-day; and of whom it shall be said in the eternal state of unclouded brightness, where sun and moon are no more, "The glory of the Lord shall lighten it, and the Lamb is the light thereof."

But suppose matter could be dowered, that worlds could have a gravitation, one of two things must follow: It must have conscious knowledge of the position, exact weight, and distance of every atom, mass, and world, in order to proportion the exact amount of gravity, or it must fill infinity with an omnipresent attractive power, pulling in myriads of places at nothing; in a few places at worlds. Every world must exert an infinitely extended power, but myriads of infinities cannot be in the same space. The solution is, one infinite power and conscious will.

To see the impossibility of every other solution, join in the long and microscopic hunt for the ultimate particle, the atom; and if found, or if not found, to a consideration of its remarkable powers. Bring telescopes and microscopes, use all strategy, for that atom is difficult to catch. Make the first search with the microscope: we can count 112,000 lines ruled on a glass plate inside of an inch. But we are here looking at mountain ridges and valleys, not atoms. Gold can be beaten to the $\frac{1}{340000}$ of an inch. It can be drawn as the coating of a wire a thousand times thinner, to the $\frac{1}{340000000}$ of an inch. But the atoms are still heaped one upon another.

Take some of the infusorial animals. Alonzo Gray says millions of them would not equal in bulk a grain of sand. Yet each of them performs the functions of respiration, circulation, digestion, and locomotion. Some of our blood-vessels are not a millionth of our size. What must be the size of the ultimate particles that freely move about to nourish an animal whose totality is too small to estimate? A grain of musk gives off atoms enough to scent every part of the air of a room. You detect it above, below, on every side. Then let the zephyrs of summer and the blasts of winter sweep through that room for forty years, bearing out into the wide world miles on miles of air, all perfumed from the atoms of that grain of musk, and at the end of the forty years the weight of musk has not appreciably dimin-

ished.   Yet uncountable myriads on myriads of atoms
have gone.

Our atom is not found yet.   Many are the ways of
searching for it which we cannot stop to consider.   We
will pass in review the properties with which materi-
alists preposterously endow it.   It is impenetrable and
indivisible.   Atoms of arsenic and phosphorus are one-
half; and of mercury and zinc, twice the normal size.
They have different shapes.   They differ in weight, in
quantity of combining power, in quality of combining
power.   They combine with different substances, in certain exact assignable quantities.   Thus, one atom of bro-
mine combines with one of hydrogen, one of oxygen with
two of hydrogen, one of nitrogen with three of hydrogen,
one of silicon with four of hydrogen, etc.   Hence our
atom of hydrogen must have power to count, or at least
to measure, or be cognizant of bulk.   Again, atoms are
of different sorts, as positive or negative to electric
currents.   They have power to take different shapes
with different atoms in crystallization; that is, there is
a power in them, conscious or otherwise, that the same
bricks shall make themselves into stables or palaces, sew-
ers or pavements, according as the mortar varies.   "No,
no," you cry out; "it is only according as the builder
varies his plan."   There is no need to rehearse these
powers much further; though not one-tenth of the
supposed innate properties of this infinitesimal infinite
have been recited—properties which are expressed by the
words atomicity, quantivalence, monad, dyad, univalent,
perissad, quadrivalent, and twenty other terms, each ex-
pressing some endowment of power in this invisible
atom.   Refer to one more presumed ability, an ability

to keep themselves in exact relation of distance and power to each other, without touching.

It is well known that water does not fill the space it occupies. We can put eight or ten similar bulks of different substances into a glass of water without greatly increasing its bulk, some actually diminishing it. A philosopher has said that the atoms of oxygen and hydrogen are probably not nearer to each other in water than one hundred and fifty men would be if scattered over the surface of England, one man to four hundred square miles.

The atoms of the luminiferous ether are infinitely more diffused, and yet its interactive atoms can give 577 millions of millions of light-waves a second. And now, more preposterous than all, each atom has an attractive power for every other atom of the universe. The little mote, visible only in a sunbeam streaming through a dark room, and the atom, infinitely smaller, has a grasp upon the whole world, the far-off sun, and the stars that people infinite space. The Sage of Concord advises you to hitch your wagon to a star. But this is hitching all stars to an infinitesimal part of a wagon. Such an atom, so dowered, so infinite, so conscious, is an impossible conception.

But if matter could be so dowered as to produce such results by mechanism, could it be dowered to produce the results of intelligence? Could it be dowered with power of choice without becoming mind? If oxygen and hydrogen could be made able to combine into water, could the same unformed matter produce in one case a plant, in another a bird, in a third a man; and in each of these put bone, brain, blood, and nerve in proper relations? Matter must be mind, or subject to a present working mind, to do this. There must be a present intelligence directing the process, laying the dead bricks, marble, and wood in an intelligent order for a living temple. If we do put God behind a single veil in dead matter, in all living things he must be apparent and at work. If, then, such a thing as an infinite atom is impossible, shall we not best understand matter by saying it is a visible representation of God's personal will and power, of his personal force, and perhaps knowledge, set aside a little from himself, still possessed somewhat of his personal attributes, still responsive to his will. What we call matter may be best understood as God's force, will, knowledge, rendered apparent, static, and unweariably operative. Unless matter is eternal, which is unthinkable, there was nothing out of which the world could be made, but God himself; and, reverently be it said, matter seems to retain fit capabilities for such source. Is not this the teaching of the Bible? I come to the old Book. I come to that man who was taken up into the arcana of the third heaven, the holy of holies, and heard things impossible to word. I find he makes a clear, unequivocal statement of this truth as God's revelation to him. "By faith," says the author of Hebrews, "we understand the worlds were framed by the word of God, so that things which are seen were not made of things which do appear." In Corinthians, Paul says—But to us there is but one God, the Father, of whom [as a source] are all things; and one Lord Jesus Christ, by whom [as a creative worker] are all things. So in Romans he says—"For out of him, and through him, and to him are all things, to whom be glory forever. Amen."

God's intimate relation to matter is explained. No wonder the forces respond to his will; no wonder pantheism—the idea that matter is God—has had such a hold upon the minds of men. Matter, derived from him, bears marks of its parentage, is sustained by him, and when the Divine will shall draw it nearer to himself the new power and capabilities of a new creation shall appear. Let us pay a higher respect to the attractions and affinities; to the plan and power of growth; to the wisdom of the ant; the geometry of the bee; the migrating instinct that rises and stretches its wings toward a provided South—for it is all God's present wisdom and power. Let us come to that true insight of the old prophets, who are fittingly called seers; whose eyes pierced the veil of matter, and saw God clothing the grass of the field, feeding the sparrows, giving snow like wool and scattering hoar-frost like ashes, and ever standing on the bow of our wide-sailing world, and ever saying to all tumultuous forces, "Peace, be still." Let us, with more reverent step, walk the leafy solitudes, and say:

> "Father, thy hand
> Hath reared these venerable columns. Thou
> Did'st weave this verdant roof. Thou did'st look down
> Upon the naked earth, and forthwise rose
> All these fair ranks of trees. They in Thy sun
> Budded, and shook their green leaves in Thy breeze.

> "That delicate forest flower,
> With scented breath and looks so like a smile,
> Seems, as it issues from the shapeless mould,
> An emanation of the indwelling life,
> A visible token of the unfolding love
> That are the soul of this wide universe."—BRYANT.

Philosophy has seen the vast machine of the universe, wheel within wheel, in countless numbers and hopeless intricacy. But it has not had the spiritual insight of Ezekiel to see that they were every one of them full of eyes—God's own emblem of the omniscient supervision.

What if there are some sounds that do not seem to be musically rhythmic. I have seen where an avalanche broke from the mountain side and buried a hapless city; have seen the face of a cliff shattered to fragments by the weight of its superincumbent mass, or pierced by the fingers of the frost and torn away. All these thunder down the valley and are pulverized to sand. Is this music? No, but it is a tuning of instruments. The rootlets seize the sand and turn it to soil, to woody fibre, leafy verdure, blooming flowers, and delicious fruit. This asks life to come, partake, and be made strong. The grass gives itself to all flesh, the insect grows to feed the bird, the bird to nourish the animal, the animal to develop the man.

Notwithstanding the tendency of all high-class energy to deteriorate, to find equilibrium, and so be strengthless and dead, there is, somehow, in nature a tremendous push upward. Ask any philosopher, and he will tell you that the tendency of all endowed forces is to find their equilibrium and be at rest — that is, dead. He draws a dismal picture of the time when the sun shall be burned out, and the world float like a charnel ship through the dark, cold voids of space—the sun a burned-out char, a dead cinder, and the world one dismal silence, cold beyond measure, and dead beyond consciousness. The philosopher has wailed a dirge without hope, a requiem without grandeur, over the world's future. But nature herself, to all ears attuned, sings pæans, and shouts to men that the highest energy, that of life, does not deteriorate.

Mere nature may deteriorate. The endowments of force must spend themselves. Wound-up watches and worlds must run down. But nature sustained by unexpendable forces must abide. Nature filled with unexpendable forces continues in form. Nature impelled by a magnificent push of life must ever rise.

Study her history in the past. Sulphurous realms of deadly gases become solid worlds; surplus sunlight becomes coal, which is reserved power; surplus carbon becomes diamonds; sediments settle until the heavens are azure, the air pure, the water translucent. If that is the progress of the past, why should it deteriorate in the future?

There is a system of laws in the universe in which the higher have mastery over the lower. Lower powers are constitutionally arranged to be overcome; higher powers are constitutionally arranged for mastery. At one time the water lies in even layers near the ocean's bed, in obedience to the law or power of gravitation. At another time it is heaved into mountain billows by the shoulders of the wind. Again it flies aloft in the rising mists of the morning, transfigured by a thousand rainbows by the higher powers of the sun. Again it develops the enormous force of steam by the power of heat. Again it divides into two light flying airs by electricity. Again it stands upright as a heap by the power of some law in the spirit realm, whose mode of working we are not yet large enough to comprehend. The water is solid, liquid, gaseous on earth, and in air according to the grade of power operating upon it.

The constant invention of man finds higher and higher powers. Once he throttled his game, and often perished in the desperate struggle; then he trapped it; then pierced it with the javelin; then shot it with an arrow, or set the springy gases to hurl a rifle-ball at it. Sometime he may point at it an electric spark, and it shall be his. Once he wearily trudged his twenty miles a day, then he took the horse into service and made sixty; invoked the winds, and rode on their steady wings two hundred and forty; tamed the steam, and made almost one thousand; and if he cannot yet send his body, he can his mind, one thousand miles a second. It all depends upon the grade of power he uses. Now, hear the grand truth of nature: as the years progress the higher grades of power increase. Either by discovery or creation, there are still higher class forces to be made available. Once there was no air, no usable electricity. There is no lack of those higher powers now. The higher we go the more of them we find. Mr. Lockyer says that the past ten years have been years of revelation concerning the sun. A man could not read in ten years the library of books created in that time concerning the sun. But though we have solved certain problems and mysteries, the mysteries have increased tenfold.

We do not know that any new and higher forces have been added to matter since man's acquaintance with it. But it would be easy to add any number of them, or change any lower into higher. That is the meaning of the falling granite that becomes soil, of the pulverized lava that decks the volcano's trembling sides with flowers; that is the meaning of the grass becoming flesh, and of all high forces constitutionally arranged for mastery over lower. Take the ore from the mountain. It is loose, friable, worthless in itself. Raise it in capacity to cast-iron, wrought-iron, steel, it becomes a highway for the commerce of nations, over the mountains and under them. It becomes bones, muscles, body for the inspiring soul of steam. It holds up the airy bridge over the deep chasm. It is obedient in your hand as blade, hammer, bar, or spring. It is inspirable by electricity, and bears human hopes, fears, and loves in its own bosom. It has been raised from valueless ore. Change it again to something as far above steel as that is above ore. Change all earthly ores to highest possibility; string them to finest tissues, and the new result may fit God's hand as tools, and thrill with his wisdom and creative processes, a body fitted for God's spirit as well as the steel is fitted to your hand. From this world take opacity, gravity, darkness, bring in more mind, love, and God, and then we will have heaven. An immanent God makes a plastic world.

When man shall have mastered the forces that now exist, the original Creator and Sustainer will say, "Behold, I create all things new." Nature shall be called nearer to God, be more full of his power. To the long-wandering Æneas, his divine mother sometimes came to cheer his heart and to direct his steps. But the goddess only showed herself divine by her departure; only when he stood in desolation did the hero know he had stood face to face with divine power, beauty, and love. Not so the Christian scholars, the wanderers in Nature's bowers to-day. In the first dawn of discovery, we see her full of beauty and strength; in closer communion, we find her full of wisdom; to our perfect knowledge, she reveals an indwelling God in her; to our ardent love, she reveals an indwelling God in us.

But the evidence of the progressive refinements of habitation is no more clear than that of progressive refinement of the inhabitant: there must be some one to use these finer things. An empty house is not God's ideal nor man's. The child may handle a toy, but a man must mount a locomotive; and before there can be New Jerusalems with golden streets, there must be men more avaricious of knowledge than of gold, or they would dig them up; more zealous for love than jewels, or they would unhang the pearly gates. The uplifting refinement of the material world has been kept back until there should appear masterful spirits able to handle the higher forces. Doors have opened on every side to new realms of power, when men have been able to wield them. If men lose that ability they close again, and shut out the knowledge and light. Then ages, dark and feeble, follow.

Some explore prophecy for the date of the grand transformation of matter by the coming of the Son of Man, for a new creation. A little study of nature would show that the date cannot be fixed. A little study of Peter would show the same thing. He says, "What manner of persons ought ye to be, in all holy conversation and godliness, looking for and hastening the coming of the day of God, wherein the heavens being on fire shall be dissolved, and the elements shall melt with fervent heat? Nevertheless we, according to his promise, look for a new heaven and a new earth."

The idea is, that the grand transformation of matter waits the readiness of man. The kingdom waits the king. The scattered cantons of Italy were only prostrate provinces till Victor Emanuel came, then they were developed into united Italy. The prostrate provinces of matter are not developed until the man is victor, able to rule there a realm equal to ten cities here. Every good man hastens the coming of the day of God and nature's renovation. Not only does inference teach that there must be finer men, but fact affirms that transformation has already taken place. Life is meant to have power over chemical forces. It separates carbon from its compounds and builds a tree, separates the elements and builds the body, holds them separate until life withdraws. More life means higher being. Certainly men can be refined and recapacitated as well as ore. In Ovid's "Metamorphoses" he represents the lion in process of formation from earth, hind quarters still clay, but fore quarters, head, erect mane, and blazing eye—live lion—and pawing to get free. We have seen winged spirits yet linked to forms of clay, but beating the celestial air, endeavoring to be free; and we have seen them, dowered with new sight, filled with new love, break loose and rise to higher being.

In this grand apotheosis of man which nature teaches, progress has already been made. Man has already outgrown his harmony with the environment of mere matter. He has given his hand to science, and been lifted up above the earth into the voids of infinite space. He

has gone on and on, till thought, wearied amidst the infinities of velocity and distance, has ceased to note them.
But he is not content; all his faculties are not filled.
He feels that his future self is in danger of not being
satisfied with space, and worlds, and all mental delights,
even as his manhood fails to be satisfied with the ma-
teriel toys of his babyhood.   He asks for an Author and
Maker of things, infinitely above them.   He has seen
wisdom unsearchable, power illimitable; but he asks for
personal sympathy and love.   Paul expresses his feeling: every creature—not the whole creation—groaneth
and travaileth in pain together until now, waiting for
the adoption—the uplifting from orphanage to parentage—a translation out of darkness into the kingdom
of God's dear Son.   He hears that a man in Christ
is a new creation: old things pass away, all things become new.   There is then a possibility of finding the
Author of nature, and the Father of man.   He begins
his studies anew.   Now he sees that all lines of knowledge converge as they go out toward the infinite mystery; sees that these converging lines are the reins of
government in this world; sees the converging lines
grasped by an almighty hand; sees a loving face and
form behind; sees that these lines of knowledge and
power are his personal nerves, along which flashes his
will, and every force in the universe answers like a
perfect muscle.

Then he asks if this Personality is as full of love as
of power.   He is told of a tenderness too deep for tears,
a love that has the Cross for its symbol, and a dying
cry for its expression: seeking it, he is a new creation.
He sees more wondrous things in the Word than in the

world. He comes to know God with his heart, better than he knows God's works by his mind.

Every song closes with the key-note with which it began, and the brief cadence at the close hints the realms of sound through which it has tried its wings. The brief cadence at the close is this: All force runs back into mind for its source, constant support, and uplifts into higher grades.

Mr. Grove says, " Causation is the will, creation is the act, of God." Creation is planned and inspired for the attainment of constantly rising results. The order is chaos, light, worlds, vegetable forms, animal life, then man. There is no reason to pause here. This is not perfection, not even perpetuity. Original plans are not accomplished, nor original force exhausted. In another world, free from sickness, sorrow, pain, and death, perfection of abode is offered. Perfection of inhabitant is necessary; and as the creative power is everywhere present for the various uplifts and refinements of matter, it is everywhere present with appropriate power for the uplifting and refinement of mind and spirit.

SUMMARY OF LATEST DISCOVERIES AND CONCLUSIONS.

Movements on the Sun.—The discovery and measurement of the up-rush, down-rush, and whirl of currents about the sun-spots, also of the determination of the velocity of rotation by means of the spectroscope, as described (page 53), is one of the most delicate and difficult achievements of modern science.

Movement of Stars in Line of Sight (page 51).—The following table shows this movement of stars, so far as at present known:

APPROACHING.			RECEDING.		
Map.	Name.	Rate per sec.	Map.	Name.	Rate per sec.
Fig. 71	Arcturus.....	55 miles	Fig. 69	Sirius	20 miles
" 72	Vega..........	50 "	Fr'piece	Betelguese...	22 "
" 73	a Cygni	39 "	"	Rigel..........	15 "
" 69	Pollux........	49 "	Fig. 69	Castor.........	25 "
" 67	Dubhe........	46 "	" 70	Regulus.......	15 "

Sun's Appearance.—This was formerly supposed to be an even, regular, dazzling brightness, except where the spots appeared. But the sun's surface is now known to be mottled with what are called rice grains or willow leaves. But the rice grains are as large as the continent of America. The spaces between are called pores. They constitute an innumerable number of small spots. This appearance of the general surface is well portrayed in the cut on page 92.

Close Relation between Sun and Earth.—Men always knew that the earth received light from the sun. They subsequently discovered that the earth was momentarily held by the power of gravitation. But it is a recent discovery that the light is one of the principal agents in chemical changes, in molecular grouping and world-building, thus making all kinds of life possible (p. 30–36). The close connection of the sun and the earth will be still farther shown in the relation of sun-spots and auroras. One of the most significant instances is related on page 19, when the earth felt the fall of bolides upon the sun. Members of the body no more answer to the heart than the planets do to the sun.

Hydrogen Flames.—It has been demonstrated that the sun flames 200,000 miles high are hydrogen in a state of flaming incandescence (page 85).

Sun's Distance.—The former estimate, 95,513,794 miles, has been reduced by nearly one-thirtieth. Lockyer has stated it as low as 89,895,000 miles, and Proctor, in "Encyclopædia Britannica," at 91,430,000 miles, but discovered errors show that these estimates are too small. Newcomb gives 92,400,000 as within 200,000 miles of the correct distance. The data for a new determination of this distance, obtained from the transit of Venus, December 8th, 1874, have not yet been deciphered; a fact that shows the difficulty and laboriousness of the work. Meanwhile it begins to be evident that observations of the transit of Venus do not afford the best basis for the most perfect determination of the sun's distance.

Since the earth's distance is our astronomical unit of measure, it follows that all other distances will be changed, when expressed in miles, by this ascertained change of the value of the standard.

Oxygen in the Sun.—In 1877 Professor Draper announced the discovery of oxygen lines in the spectrum of the sun. The discovery was doubted, and the methods used were criticised by Lockyer and others, but later and more delicate experiments substantiate Professor Draper's claim to the discovery. The elements known to exist in the sun are salt, iron, hydrogen, magnesium, barium, copper, zinc, cromium, and nickel. Some elements in the sun are scarcely, if at all, discoverable on the earth, and some on the earth not yet discernible in the sun.

Substance of Stars.—Aldebaran (*Frontispiece*) shows salt, magnesium, hydrogen, calcium, iron, bismuth, tellurium, antimony, and mercury. Some of the sun's metals do not appear. Stars differ in their very substance, and will, no doubt, introduce new elements to us unknown before.

The theory that all nebulæ are very distant clusters of stars is utterly disproved by the clearest proof that some of them are only incandescent gases of one or two kinds.

Discoveries of New Bodies.— Comets. The companion of Sirius (p. 211). The two satellites of Mars were discovered by Mr. Hall, U. S. Naval Observatory, August 11th, 1877 (page 161). The outer one is called Diemos; the inner, Phobos.

Sir William Herschel thought he discovered six satellites of Uranus. The existence of four of them has been disproved by the researches of men with larger telescopes. Two new ones, however, were discovered by Mr. Lassell in 1846.

Saturn's Rings are proved to be in a state of fluidity and contraction (page 171).

Meteors and Comets.— The orbits of over one hundred swarms of meteoric bodies are fixed: their relation to, and in some cases indentity with, comets determined. Some comets are proved to be masses of great weight and solidity (page 133).

Aerolites.— Some have a texture like our lowest strata of rocks. There is a geology of stars and meteors as well as of the earth. M. Meunier has just received the Lalande Medal from the Paris Academy for his treatise showing that, so far as our present knowledge can determine, some of these meteors once belonged to a globe developed in true geological epochs, and which has been separated into fragments by agencies with which we are not acquainted.

Dr. O. Hahn, a German lawyer, proves the existence of

organic remains in the stones that come from space. From museums in Tubingen and in Vienna Hahn procured himself more than six hundred chips of meteorites of the Choadrite class, proved in each case to be genuine, and having been collected on eighteen different occasions, partly during the present and partly during the last century, in Europe, Asia, and America. Minute inspection has discovered in them a quantity of organic remains, principally belonging to the most ancient form of porous corallines, to the genus of fossil zoophytes denominated Favosites, or at least bearing a very strong resemblance to these latter, though of a still smaller type. About fifty kinds of these tiny animals have been made out by Dr. Hahn, and assigned to sixteen different families.

Dr. D. F. Weinland, who has devoted a year to the study of Hahn's book, and testing his microscopic researches, says that it is only the shell of the Choadrite meteorite that is burnt and glazed by friction with our atmosphere. The heat does not extend so far during the short transit of the meteor as to impair the kernel, which has an appearance somewhat like coarse shell lime, of a conglomeration of petrified organic matter, baked in a lump. Though only few specimens can be called well preserved, yet the substance is sufficiently distinguishable to enable us to class most of the structures among the Polycistines and the Foraminifera. They must have existed in water warm enough never to freeze down to the bottom. Where are we to seek for this water, if Professor Schiaparelli tells us that meteorites do not belong to our solar system, but are intruders from without? Very strange is the complete resemblance of all the cuttings examined to one another, though, as stated, they belong to stones fallen at different periods in all parts of the globe. Are these parts of an exploded world, or have these little worlds developed life in a manner similar to larger ones?

The Horizontal Pendulum.—This delicate·instrument is represented in Fig. 82. It consists of an upright standard, strongly braced; a weight, *m*, suspended by the hair-spring of a watch, B D, and held in a horizontal position by another watch-spring, A C. The weight is deflected from side to side by the slightest influence. The least change in the level of a base thirty-nine inches long that could be detected by a spirit-level is $0''.1$ of an arc—equal to raising one end $\frac{1}{10000}$ of an inch. But the

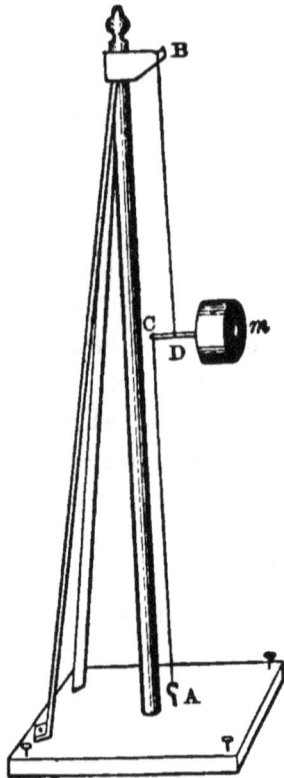

Fig. 82.—Horizontal Pendulum.

pendulum detects a raising of one end $\frac{1}{38000000}$ of an inch. To observe the movements of the pendulum, it is kept in a dark room, and a ray of light is directed to the mirror, *m*, and thence reflected upon a screen. Thus the least movement may be enormously magnified, and read and measured by the moving spot on the screen. It has been discovered that when the sun rises it has sufficient attraction to incline this instrument to the east; when it sets, to incline it to the west. The same is true of the moon. When either is exactly overhead or underfoot, of course there is no deflection. The mean deflection caused by the moon at rising or setting is $0''.0174$; by the sun, $0''.008$. Great results are expected from this instrument hardly known as yet: among others, whether gravitation acts instantly or consumes time in coming from the sun. This will be shown by the time of the change of the pendulum from east to west when the sun reaches the zenith, and *vice versa* when it crosses the nadir. The sun will be best studied without light, in the quiet and darkness of some deep mine.

Light of Unseen Stars.—From careful examination, it appears that three-fourths of the light on a fine starlight night comes from stars that cannot be discerned by the naked eye. The whole amount of star light is about one-eightieth of that of the full moon.

Lateral Movements of Stars, page 226–28.

Future Discoveries—A Trans-Neptunian Planet.—Professor Asaph Hall says: "It is known to me that at least two American astronomers, armed with powerful telescopes, have been searching quite recently for a trans-Neptunian planet. These searches have been caused by the fact that Professor Newcomb's tables of Uranus and Neptune already begin to differ from observation. But are we to infer from these errors of the planetary tables the existence of a trans-Neptunian planet? It is possible that such a planet may exist, but the probability is, I think, that the differences are caused by errors in the theories of these planets. * * * A few years ago the remark was frequently made that the labors of astronomers on the solar system were finished, and that henceforth they could turn their whole attention to sidereal astronomy. But to-day we have the lunar theory in a very discouraging condition, and the theories of Mercury, Jupiter, Saturn, Uranus, and Neptune all in need of revision; unless, indeed, Leverrier's theories of the last two planets shall stand the test of observation. But, after all, such a condition of things is only the natural result of long and accurate series of observations, which make evident the small inequalities in the motions, and bring to light the errors of theory."

Future discoveries will mostly reveal the laws and conditions of the higher and finer forces. Already Professor Loomis telegraphs twenty miles without wire, by the electric currents between mountains. We begin to use electricity for light, and feel after it for a motor. Comets and Auroras show its presence between worlds, and in the interstellar spaces. Let another Newton arise.

12*

SOME ELEMENTS OF THE SOLAR SYSTEM.

Name.	Sign.	Masses.	Mean Dist. from Sun. Earth's Dist. as 1.	Mean Dist. from Sun. Millions of Miles.	Mean Diameter in Miles.	Density. ⊕ = 1.	Axial Revolution.	Gravity at Surface. ⊕ = 1.	Periodic Time.	Orbital Velocity in Miles per Sec.
Sun............	☉	Unity.	......	......	860,000	0.255	25 to 26^d	27.71	......	......
Mercury........	☿	$\frac{1}{30000000}$ (?)	0.387	35¾	2,992	1.21	24^h 5^m (?)	0.46	87.97^d	29.55
Venus..........	♀	$\frac{1}{425000}$	0.723	66¾	7,660	0.85	23^h 21^m (?)	0.82	224.70^d	21.61
Earth..........	⊕	$\frac{1}{355000}$	1.	92¾	7,918	1.	23^h 56^m 4^s	1.	365.26^d	18.38
Mars...........	♂	$\frac{1}{3930000}$	1.523	141	4,211	0.737	24^h 37^m 22.7^s	0.39	686.98^d	14.99
Asteroids.......	(No.)	......	......	......	......	......	......	......	......	......
Jupiter.........	♃	$\frac{1}{1047}$	5.203	480	86,000	0.243	9^h 55^m 20^s	2.64	11.86^{yrs}	8.06
Saturn..........	♄	$\frac{1}{3501}$	9.538	881	70,500	0.133	10^h 14^m	1.18	29.46^{yrs}	5.95
Uranus..........	♅	$\frac{1}{22600}$	19.188	1771	31,700	0.226	Unknown.	0.90	84.02^{yrs}	4.20
Neptune.........	♆	$\frac{1}{19380}$	30.054	2775	34,500	0.204	Unknown.	0.89	164.78^{yrs}	3.86

EXPLANATION OF ASTRONOMICAL SYMBOLS.

SIGNS OF THE ZODIAC.

0. ♈ Aries...............	0°	VI. ♎ Libra...............	180°
I. ♉ Taurus...............	30	VII. ♏ Scorpio...............	210
II. ♊ Gemini...............	60	VIII. ♐ Sagittarius............	240
III. ♋ Cancer...............	90	IX. ♑ Capricornus.........	270
IV. ♌ Leo...............	120	X. ♒ Aquarius............	300
V. ♍ Virgo...............	150	XI. ♓ Pisces...............	330

☌ Conjunction.	S. Seconds of Time.
☐ Quadrature.	° Degrees.
☍ Opposition.	' Minutes of Arc.
☊ Ascending Node.	" Seconds of Arc.
☋ Descending Node.	R. A. Right Ascension.
H. Hours.	Decl. or D. Declination.
M. Minutes of Time.	N. P. D. Dist. from North Pole.

OTHER ABBREVIATIONS USED IN THE ALMANAC.

S., south, *i.e.*, crosses the meridian; M., morning; A, afternoon; Gr. H. L. N., greatest heliocentric latitude north, *i.e.*, greatest distance north of the ecliptic, as seen from the sun. ☌ ☿ ☉ Inf., inferior conjunction; Sup., superior conjunction.

GREEK ALPHABET USED INDICATING THE STARS.

α, alpha.	η, eta.	ν, nu.	τ, tau.
β, beta.	θ, theta.	ξ, xi.	υ, upsilon.
γ, gamma.	ι, iota.	o, omicron.	φ, phi.
δ, delta.	κ, kappa.	π, pi.	χ, chi.
ε, epsilon.	λ, lambda.	ρ, rho.	ψ, psi.
ζ, zeta.	μ, mu.	σ, sigma.	ω, omega.

CHAUTAUQUA OUTLINE FOR STUDENTS.

As an aid to comprehension, every student should draw illustrative figures of the various circles, planes, and situations described. (For example, see Fig. 45, page 112.) As an aid to memory, the portion of this outline referring to each chapter should be examined at the close of the reading, and this mere sketch filled up to a perfect picture from recollection.

I. *Creative Processes.*—The dial-plate of the sky. Cause of different weights—on sun, moon. Two laws of gravity. Inertia. Fall of earth to sun per second. Forward motion. Elastic attraction. Perturbation of moon; of Jupiter and Saturn. Oscillations of planets.

II. *Light.*—From condensation. Number of vibrations of red; violet. Thermometer against air. Aerolite against earth. Two bolides against the sun. Large eye. Velocity of light. Prism. Color means different vibrations. Music of light. Light reports substance of stars. Force of; bridge, rain, dispersion, intensities, reflection, refraction, decomposition.

III. *Astronomical Instruments.* — Refracting telescope. Reflecting; largest. Spectroscope. Spectra of sun, hydrogen, sodium, etc. E made G by approach; C by departure. Stars approach and recede.

IV. *Celestial Measurements.*—Place and time by stars. Degrees, minutes, seconds. Mapping stars. Mural circle. Slow watch. Hoosac Tunnel. Fine measurements. Sidereal time. Spider-lines. Personal equation. Measure distance—height. Ten-inch base line. Parallax of sun, stars. Longitude at sea. Distance of Polaris, *a* Centauri, 61 Cygni. Orbits of asteroids.

V. *The Sun.*—World on fire. Apparent size from planets. Zodiacal light. Corona. Hydrogen — how high? Size. How many earths? Spots: 1. Motion; 2. Edges; 3. Variable; 4. Periodic; 5. Cyclonic; 6. Size; 7. Velocities. What the sun does. Experiments.

VI. *The Planets from Space.*—North Pole. Speed. Sizes. Axial revolution. Man's weight on. Seasons. Parallelism of axis. Earth near sun in winter. Plane of ecliptic. Orbits inclined to. Earth rotates. Proof. Sun's path among stars. Position of planets. Motion—direct, retrograde. Experiments.

VII. *Meteors.*—Size; number; cause of; above earth; velocity; colors; number in space; telescopic view of. Aerolites: Systems of; how many known. Comets: Orbits; number of comets; Halley's; Biela's lost; Encke's. Resisting medium. Whence come comets? Composed of what? Amount of matter in. ⊕.

VIII. *The Planets.*—How many? Uranus discovered? Neptune? Asteroids? Vulcan? Distance from sun. Periodic time. Mercury: Elements; shapes, as seen from earth; transits. Venus: Elements; seen by day; how near earth? how far from? phases; Galileo. Earth: Elements; in space; Aurora; balance of forces. Tides: Main and subsidiary causes; eastern shores; Mediterranean Sea. Moon: Elements; hoax; moves east; see one side; three causes help to see more than half. Revolution: Why twenty-nine and a half days: heat—cold; how much light? Craters and peaks lighted; measured. Eclipses—Why not every new and full moon? Periodicity. Mars: Elements; how near earth? How far from? Apparent size; ice-fields; which end most? Satellites—Asteroids: How found? When? By whom? How many? Jupiter: Elements; trade-winds; how much light received? Own heat. Satellites: How many? Colors. Saturn: Elements; habitability; rings; flux; satellites. Uranus: Elements; discoverer; seen by; moon's motion. Neptune: Elements; discovered by; how? Review system.

IX. *The Nebular Hypothesis.*—State it; facts confirmatory. Objections—1. Heat; 2. Rotation; 3. Retrograde; 4. Martial moons; 5. Star of 1876. Evolution: Gaps in; conclusion.

X. *The Stellar System.*—Motto. Man among stars; open page; starry poem; stars located; named. Thuban. Etanin. Constellations: Know them; number of stars; double; ε Lyræ, Sirius, Procyon, Castor, 61 Cygni, γ Virginis. Colored stars; change color. Clusters: Two theories. Nebulæ: Two visible; composed of; shapes; where? Variable stars. Sun. β Lyræ, Mira, Betelguese, Algol; cause. Temporary; 1572. New star of 1866: Two theories. Star of 1876. Movements of stars; Sirius; sun; 1830 Groombridge. Stars near Pleiades: Orion, Great Dipper, Southern Cross. Centre of gravity.

XI. *The Worlds and the Word.*—Rich. Number. Erroneous allusions. Truth before discovery: 1. A beginning; 2. Creation before arrangement; 3. Light before sun; 4. Mountains under water; 5. Order of development;

6. Sphere of earth; 7. How upheld; 8. Number of stars; 9. Weight of air; 10. Meteorology; 11. Queries to Job; 12. Sun to end of heaven; 13. View of Mitchell; 14. Herschel. What is matter? Force? End of earth. Way to knowledge. Work of light. Transfiguration of matter. Uniformitarianism. A whisper of Him. Man for mastery. Each a type of higher. Survival of fittest. Uranus. Worlds and Word one language.

XII. *The Ultimate Force.*—Universe shows power: 1. Rain; Niagara; 2. Vegetable growth; 3. Worlds carried; 4. Sun; fill dome with worlds; 5. Double suns; 6. Galaxies. Correlation. What ultimate? Mind and will. What continuous relation? Watch. Theories of gravitation: Newton's, Le Sage's, Bible's. High-class energy deteriorates. Search for atoms: 1. Microscope; 2. Gold; 3. Infusoria; 4. Musk. Properties of atoms: 1. Impenetrable; 2. Indivisible; 3. Shape; 4. Quality; 5. Crystallization; 6. Not touch each other; 7. Active; 8. Attractive; 9. Intelligent. Whose? Relation of matter to God; rock to soil. Push upward. Highest has mastery. Man advances by highest. Matter recapacitated. Refined habitations. Inhabitants. All force leads back to mind. Personal and infinite.

GLOSSARY OF ASTRONOMICAL TERMS AND INDEX.

Abbreviations used in astronomies, 275.

Aberration of light (*a wandering away*), an apparent displacement of a star, owing to the progressive motion of light combined with that of the earth in its orbit, 199.

Aerolite (*air-stone*), 122.

Air, refraction of the, 40.

Algol, the variable star, 222.

Almanac, Nautical, 71; explanation of signs used, 275.

Alphabet, Greek, 275.

Altitude, angular elevation of a body above the horizon.

Angle, difference in directions of two straight lines that meet.

Annular (*ring-shaped*) eclipses, 158; nebulæ, 218, 220.

Aphelion, the point in an orbit farthest from the sun.

Apogee, the point of an orbit which is farthest from the earth.

Apsis, plural *apsides*, the line joining the aphelion and perihelion points; or the major axis of elliptical orbits.

Arc, a part of a circle.

Ascension, right, the angular distance of a heavenly body from the first point of Aries, measured on the equator.

Asteroids (*star-like*), 162; orbits of interlaced, 74.

Astronomical instruments, 43.

Astronomy, use of, 57.

Atom, size of, 255; power of, 256.

Aurora Borealis, 143.

Axis, the line about which a body rotates.

Azimuth, the angular distance of any point or body in the horizon from the north or south points.

Bailey's beads, dots of light on the edge of the moon seen in a solar eclipse, caused by the moon's inequalities of surface.

Base line, 68.

Biela's comet, 129.

Binary system, a double star, the component parts of which revolve around their centre of gravity.

Bode's law of planetary distances is no law at all, but a study of coincidences.

Bolides, small masses of matter in space. They are usually called meteors when luminous by contact with air, 120.

Celestial sphere, the apparent dome in which the heavenly bodies seem to be set; appears to revolve, 3.

Centre of gravity, the point on which a body, or two or more related bodies, balances.

Centrifugal force (*centre fleeing*).

Chromolithic plate of spectra of metals, to face 50.

Circumpolar stars, map of north, 201.

Colors of stars, 214.

Colures, the four principal meridians of the celestial sphere passing from the pole, one through each equinox, and one through each solstice.

Comets, 126; Halley's, 128; Biela's lost, 129; Encke's, 130; constitution of, 131; will they strike the earth? 133.

Conjunction. Two or more bodies are in conjunction when they are in a straight line (disregarding inclination of orbit) with the sun. Planets nearer the sun than the earth are in inferior conjunction when they are between the earth and the sun; superior conjunction when they are beyond the sun.

Constellation, a group of stars supposed to represent some figure: circumpolar, 201; equatorial, for December, 202; for January, 203; April, 204; June, 205; September, 206; November, 207; southern circumpolar, 208.

Culmination, the passage of a heavenly body across the meridian or south point of a place; it is the highest point reached in its path.

Cusp, the extremities of the crescent form of the moon or an interior planet.

Declination, the angular distance of a celestial body north or south from the celestial equator.

Degree, the $\frac{1}{360}$ part of a circle.

Direct motion, a motion from west to east among stars.

Disk, the visible surface of sun, moon, or planets.

Distance of stars, 70.

Double stars, 210.

Earth, revolution of, 109; in space, 142; irregular figure, 145.

Eccentricity of an ellipse, the distance of either focus from centre. A planet's orbit has comparatively little, a comet's very much.

Eclipse (*a disappearance*), 157.

Ecliptic, the apparent annual path of the sun among the stars; plane of, 106.

Egress, the passing of one body off the disk of another.

Elements, the quantities which determine the motion of a planet: data for predicting astronomical phenomena; table of solar, 274.

Elements, chemical, present in the sun, 270.

Elongation, the angular distance of a planet from the sun.

Emersion, the reappearance of a body after it has been eclipsed or occulted by another.

TO FIND THE STARS IN THE SKY.

Detach any of the following maps, appropriate to the time of year, hold it between you and a lantern out-of-doors, and you have an exact miniature of the sky. Or, better, cut squares of suitable sizes from the four sides of a box; put a map over each aperture; provide for ventilation, and turn the box over a lamp or candle out-of-doors. Use an opera glass to find the smaller stars, if one is accessible.

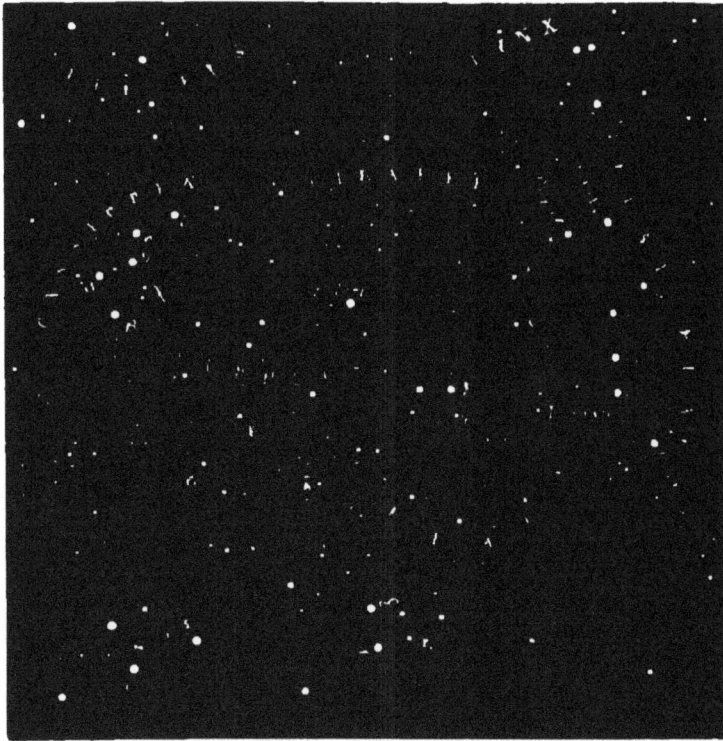

Circumpolar Constellations. Always visible. In this position.— January 20th, at 10 o'clock; Febrnary 4th, at 9 o'clock; and February 19th, at 8 o'clock.

Algol is on the Meridian, 51° South of Pole.—At 10 o'clock, December 7th; 9 o'clock December 22d; 8 o'clock, January 5th.

Capella (45° from the Pole) and Rigel (100°) are on the Meridian at 8 o'clock February 7th, 9 o'clock January 22d, and at 10 o'clock January 7th.

Regulus comes on the Meridian, 79° south from the Pole, at 10 o'clock March 23d, 9 o'clock April 8th, and at 8 o'clock April 23d.

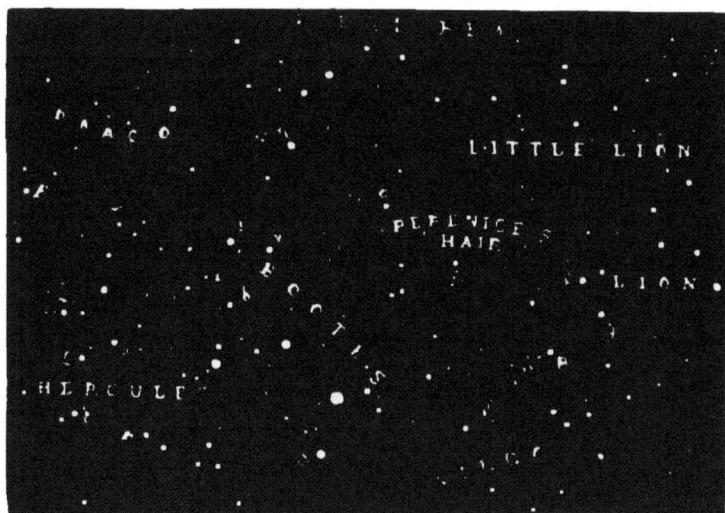

Arcturus comes to the Meridian, 70° from the Pole, at 10 o'clock May 25th, 9 o'clock June 9th, and at 8 o'clock June 25th.

13

Altair comes to the Meridian, 52° from the Pole, at 10 o'clock P.M. August 18th, at 9 o'clock September 2d, and at 8 o'clock September 18th.

Fomalhaut comes to the Meridian, only 17° from the horizon, at 8 o'clock November 4th.